PRACTICAL ECOLOGY

DIMENSIONS OF SCIENCE

DIMENSIONS OF SCIENCE
Series Editor: Professor Jeff Thompson

PRACTICAL ECOLOGY

David Slingsby
B.Sc., Ph.D., PGCE

and

Ceridwen Cook
B.Sc., PGCE

MACMILLAN

First published 1986 by
THE MACMILLAN PRESS LTD
Houndmills, Basingstoke, Hampshire RG21 2XS
and London
Companies and representatives
throughout the world

ISBN 0–333–39813–0

A catalogue record for this book is available
from the British Library.

Printed in Hong Kong

15	14	13	12	11	10	9	8	7
03	02	01	00	99	98	97	96	95

Series Standing Order

If you would like to receive future titles in this series as they are published, you can
make use of our standing order facility. To place a standing order please contact your
bookseller or, in case of difficulty,write to us at the address below with your name
and address and the name of the series. Please state with which title you wish to
begin your standing order. (If you live outside the United Kingdom we may not have
the rights for your area, in which case we will forward your order to the publisher
concerned.)

Customer Services Department, Macmillan Distribution Ltd
Houndmills, Basingstoke, Hampshire RG21 2XS, England

To all the students of Pate's and Ripon Grammar Schools who have provided the inspiration for the writing of this book, and to the many other people who have helped and encouraged us in the project and in developing our own interest in ecology

Contents

Series Editor's Preface

This book is one in a Series designed to illustrate and explore a range of ways in which scientific knowledge is generated, and techniques are developed and applied. The volumes in this Series will certainly satisfy the needs of students at 'A' level and in first-year higher-education courses, although there is no intention to bridge any apparent gap in the transfer from secondary to tertiary stages. Indeed, the notion that a scientific education is both continuous and continuing is implicit in the approach which the authors have taken.

Working from a base of 'common core' 'A'-level knowledge and principles, each book demonstrates how that knowledge and those principles can be extended in academic terms, and also how they are applied in a variety of contexts which give relevance to the study of the subject. The subject matter is developed both in depth (in intellectual terms) and in breadth (in relevance). A significant feature is the way in which each text makes explicit some aspect of the fundamental processes of science, or shows science, and scientists, 'in action'. In some cases this is made clear by highlighting the methods used by scientists in, for example, employing a systematic approach to the collection of information, or the setting up of an experiment. In other cases the treatment traces a series of related steps in the scientific process, such as investigation, hypothesising, evaluating and problem-solving. The fact that there are many dimensions to the creation of knowledge and to its application by scientists and technologists is the title and consistent theme of all the books in the Series.

The authors are all authorities in the fields in which they have written, and share a common interest in the enjoyment of their work in science. We feel sure that something of that satisfaction will be imparted to their readers in the continuing study of the subject.

Acknowledgements

The authors are grateful to the following

George Allen and Unwin, for permission to use extracts from *Seashore Studies* by M. Jenkins in sections 10.5, 10.6 and 10.7.

The Severn Trent Water Authority for permission to reproduce tables 7.5 and 7.6.

Biometrika Trustees for permission to reproduce extracts from *Biometrika*, volume 52, 1965, in table 6.2.

Thomas Nelson and Son for permission to use figures 6.1 and 6.2.

Harper & Row. Publishers, Inc., for material adapted from p. 138 in *Ecology: the Experimental Analysis of Distribution and Abundance*, Third Edition. Copyright © 1978 by Charles J. Krebs. Reprinted by permission of Harper & Row, Publishers, Inc.

Every effort has been made to trace all the copyright holders but if any have been inadvertently overlooked the publishers will be pleased to make the necessary arrangement at the first opportunity.

How to Use This Book

This book is designed to help you carry out your own field studies by describing and explaining techniques, offering suggestions and providing enough theoretical background information to enable you to organise your work and interpret your results. It is not intended to be read from cover to cover but it is suggested that you read chapters 1 and 2 before you start and carry out exercise 2.1 as an introduction to sampling. Chapter 8 also contains important principles about the relationship between organisms and their physical and chemical environment. Chapters 9 and 10 suggest how you could implement these principles either in terrestrial or seashore habitats. Chapters 6 and 7 contain a variety of exercises illustrating aspects of animal ecology, each a complete study in itself.

Other chapters are to be dipped into to 'service' chapters 9 and 10, and chapters 5, 6 and 7 also provide methods and suggestions for describing and comparing terrestrial and freshwater plant and animal communities. Chapters 3 and 4 provide methods for measuring physical and chemical factors on land, the seashore and freshwater habitats.

Chapter 11 is concerned with carrying out experiments to test hypotheses you have made during the course of your field studies and, like chapters 1, 2 and 8, contains principles which apply to whatever form your course has taken.

To aid cross-referencing, each chapter is written in numbered sections. For example, '(8.2)' refers to the second section in chapter 8. Tables, figures and exercises are numbered separately prefixed by a chapter number; for example, 'table 3.1' is the first table in chapter 3.

Note the conservation warning on page 9.

1 Ecology — Some Basic Principles

1.1 WHAT IS ECOLOGY?

Ecology is the study of plants and animals in relation to each other and to the physical and chemical environment in which they naturally occur. Much of a biology course emphasises the individual organism whether it involves dissection of a rat which has been bred in a cage, a Pelargonium taken from a greenhouse for a photosynthesis experiment or Garden Peas grown in Mendel's monastery garden. A cabbage may be homogenised just to provide chloroplasts for a biochemistry experiment or to provide material for electron microscopy. Classification involves more variety of species, but often they are dead and preserved in a jar. All these approaches to biology are important but it is also important to complete the picture by looking at the subject ecologically. Ecology means literally 'the study of the hearth' – the study of plants and animals 'at home' in the systems of which they are a part, which they rely on, help to create and maintain, and in which they have evolved.

Ecological field studies are also an important part of a biology course because they offer much scope for putting the scientific process into practice yourself rather than simply learning about it. Most aspects of biology can only be studied from books or laboratory experiments which, if correctly carried out, always produce the same result. Every site used in field studies is to some extent unique, and your observations original. You have the experience of genuinely being 'at the sharp end' of scientific research.

1.2 THE SCIENTIFIC METHOD

This is a philosophical approach to seeking after truth which involves the following aspects.

1

(a) Observation
Scientific observation involves accurate and detailed recording. It can take the form of diagrams and photographs, but often it seeks to be numerical. Observations may be made by simply using your unaided eyes, but for many you need equipment like microscopes, pH meters and light sensors.

(b) Data analysis, presentation and interpretation
Large amounts of data need to be presented in an easily interpreted form (such as a graph). It may be analysed by a branch of mathematics called statistics, in which a computer can be useful.

(c) Hypothesis
Observations alone do not prove anything, but they suggest possible explanations called hypotheses (8.2, 8.10, 11.1, exercise 11.1).

(d) Experiments
These are designed to test hypotheses (chapter 11).

(e) Reproducibility
To be scientifically acceptable, observations and experimental results should be capable of being confirmed by other people. They must always be accompanied by precise details of how they were made. Others may dispute your conclusions but they must never be able to doubt your data. Dubious results (such as those obtained by unreliable equipment) should be thrown away.

(f) Literature research
Reading about the findings and ideas of others is often essential to deciding where to start an investigation and to the interpretation of your data. In addition to using the background information in this book, you will find that access to a library is important in scientific research.

1.3 COMMUNITIES

An ecological community is an assemblage of plant, animal and microbial species in a particular place. Because different methods are involved, often plant, animal and microbial communities are considered separately. Daisy is a species frequently used as an example in this book because it is a common plant that everyone is familiar with. It is, however, only common in certain types of plant community (along with other species like Dandelion) of which it is a distinctive member, namely grazed or mown grassland. It is very rare in other communities like woods, heaths, ponds and those of the seashore. The reason certain species tend to occur together is that they share certain environmental requirements. The herbs of grassland are adapted to tolerate trampling and grazing but because the plants are small, they cannot compete with taller plants species unless mowing and grazing keeps them down.

1.4 ECOSYSTEMS

An ecosystem is an assemblage of plant, animal and microbial species in a particular place which interact with each other and with their physical and chemical environment in such a way as to constitute a self-maintaining and self-regulating system. A Beechwood represents, at one level, a community, but at a deeper one, it functions as an intricate ecosystem which can maintain itself without the help of any forester, and stays the same for thousands of years. An ecosystem has the following features.

(a) Ecosystems require an external energy source
Green plants are photosynthetic autotrophes. They absorb solar energy and use it to convert carbon dioxide into organic substances like sugars, starch, cellulose, lignin and proteins. The only substances they require for this are water and small amounts of inorganic ions like nitrate. The energy absorbed becomes incorporated into the organic molecules, and is available to the heterotrophic animals and microbes when they feed on plants (or plant-eating animals). The green plants are the primary producers of the system because they produce the energy containing organic matter which fuels the rest of it. The heterotrophes also rely on this organic matter as a source of raw materials for growth (mainly protein). Plants are also the ultimate source for mineral salts, and for many heterotrophes, most of their water (since plants, unlike animals, have roots). Herbivores (including many birds and invertebrates) represent the majority of animals and are secondary producers since they provide organic matter for carnivores. Herbivores are also regarded as primary consumers, since they feed on primary producers, and carnivores are secondary, tertiary or quaternary consumers depending on where they come in the food chain of which they are a part. All food chains (for example, grass–rabbit–fox) start ultimately with primary producers. Estuaries are rich in animal life because they receive organic matter washed down by a river as well as that derived from its own plant communities (including microscopic algae), but are often considered as 'incomplete' or 'open' ecosystems, annexes of the ecosystem upstream which provided the 'extra' organic matter. Similarly, a cave is

not a 'complete' or 'closed' ecosystem since it lacks any primary producers. The bats and the fungi which feed on their droppings are at the ends of food chains which start outside in nearby fields and woods.

Unless you consider the World as a whole, no site or even a whole continent represents an absolutely complete or closed system since all are influenced by neighbouring systems to some extent. The terms 'complete' and 'incomplete' are relative. It is convenient to regard a particular forest as a relatively complete ecosystem, but really it is but a part of the World ecosystem or biosphere. Migratory birds, for example, divide their time between this forest and another perhaps thousands of miles away.

The productivity of an ecosystem is the rate at which new organic matter (or biomass) is produced and is usually expressed as grams per unit area per unit time. Where primary productivity is high or there is an external source of organic matter, as in an estuary or a pond enriched by fallen leaves or run-off from surrounding farmland, the ecosystem supports a large number of heterotrophes. A lake in a valley composed of hard granitic rock has low productivity and consequently small animal populations because of the lack of mineral salts required by the primary producers (mainly planktonic algae).

Food chains which start with live plants are grazing foodchains.

(b) Ecosystems recycle inorganic substances

Food chains which start with dead organic matter, including animal faeces (derived originally from live plants), are detritus foodchains which include saprophytic bacteria and fungi, woodlice, earthworms and birds like thrushes (which occupy the end of the chain, feeding on invertebrates). These provide 'nature's waste-disposal system' and ultimately convert the organic matter to carbon dioxide, salts and water, and these substances are thus continually replenished or recycled. Green plants recycle oxygen, a waste product of photosynthesis.

Woodlands have a higher productivity than meadows, but much of their biomass is in the form of cellulose and lignin which most animals cannot digest. Although their grazing food chains represent few individuals, they are rich in species of fungi and other detritus feeders which are able to utilise these substances. Often these saprophytic fungi are very specialised. For example, Jew's Ear (*Auricularia auricula*) is a jelly fungus which only feeds on dead (and sometimes on living) Elder (*Sambucus niger*). Similarly, *Daldinia concentrica* feeds on Ash (*Fraxinus excelsior*). In peat bogs, acidity and anaerobic conditions greatly limit microbial activity, and much of the productivity gives rise to peat accumulation. Such places have relatively low productivity owing to poor mineral salt recycling. Carnivorous plants, such as Sundew (*Drosera rotundifolia*) are adapted to this by being able to supplement their mineral salt supply by digesting insects.

(c) Ecosystems control population sizes

All populations of organisms tend to rise because all species over-produce. One plant produces many seeds and one pair of Great Tits produces about 12 offspring in a season, and yet over a period of years, populations in a stable ecosystem remain more or less the same.

(i) Plants — as a result of competition for space and grazing by herbivores.
(ii) Herbivores — partly by competition for food but mainly by predation by carnivores.
(iii) Carnivores (and some herbivores) — partly by direct competition for food, but in many cases because their reproductive rate is reduced when food is in short supply (a kind of natural birth-control).

Basically, in a stable ecosystem, mortality is at equilibrium with natality (birth-rate) because many forms of mortality (competition for food and places to hide from predators, and disease) become more serious as the population rises, and so mortality is density-dependent. Predation is important in controlling herbivore populations because otherwise it will happen by competition for food in which case the vegetation gets over-grazed, and there is a fall in the ecosystem's productivity. In extreme cases, the vegetation dies, leading to soil erosion, and so the ecosystem has been seriously damaged and its carrying capacity has been reduced. Normally, in stable ecosystems, this never happens, or it ceases to be stable. In many places in Britain, the ecosystem still has not fully recovered from the introduction of rabbits by the Normans a mere 1000 years ago, especially as wolves and other carnivores apart from foxes have been exterminated in the meantime. Since its deliberate introduction into Britain (1952/3), myxomatosis (whose outbreaks are density-dependent) has improved the situation. A similar principle applies with animals like Robins. These are territorial, and instead of vicious competition for food decimating the prey populations, a ritualistic competition for nesting sites (in which birds do not usually get hurt) precedes breeding. Territory size is related to its productivity, and only birds with territories produce offspring.

1.5 ECOLOGICAL NICHES

A consequence of density-dependent mortality is what has been called 'the struggle for survival', which is so ruthless that only a few individuals live long enough to reproduce, passing on the genes which made them a success to the next generation. Thus it is the driving force behind evolution by Natural Selection, and it explains the remarkable way in which species are adapted to survive in the biological, physical and chemical environment in which they naturally occur. The harsh reality of Darwin's principle of 'survival of the fittest' is that unless a species has at least somewhere where it can be a winner in the struggle, it will become extinct. Two similar species must be genetically different, otherwise they would not be recognised as separate species. If environmental conditions were completely uniform throughout the World, one would be at least slightly better adapted to survive, and the other would become extinct.

The remarkable diversity of species arises because environmental conditions, as you will find in your field studies, vary from place to place, and because in any one place interspecific competition is reduced because there is more than one way of exploiting the situation. Daisies and grass may look as though they are in direct competition in a lawn, but the Daisy has deeper roots, and so it absorbs water and nutrients from a different layer of soil than the more shallow-rooted grass. A beetle in the same place does not compete with the grass for light or mineral salts. In fact, being a carnivore, it may help the grass by keeping herbovorous invertebrate populations under control. All three species share the same place, and yet occupy different ecological niches. An ecological niche is a way of exploiting an ecosystem rather than a geographical location.

The Mutual Exclusion Principle states that each niche can only be occupied by one species, and that if two species compete for the same niche, one will prove more successful and the other will fail. This implies that one habitat will include numerous ecological niches, each occupied by a different species. To understand this (very abstract) notion, you consider individual species and how they exploit the ecosystem (autoecology), and the habitat as a whole, with its patchwork of species and varying environmental factors (synecology). The Mutual Exclusion Principle is a fascinating concept, but is very slippery when it comes to defining a Niche precisely, since niches can 'overlap'. You should be able, however, to attempt to gain some understanding of the niches occupied by particular species you study, but to define them precisely you need to measure a large number of variables and plot them on a graph with a large number of axes. There are computer techniques which help to deal with this task, but it is beyond the scope of most field courses!

1.6 SUCCESSION, CLIMAX AND STABILITY

A stable ecosystem is usually the product of centuries if not millennia of development. A well-tended garden is ecologically very unstable and requires constant attention. Most of the species would not be there normally. Many are exotic, thousands of miles from the places in which they evolved and need your help in coping with the British climate, soil, pests and the species of the local flora, whose seeds keep arriving, brought by wind and animals. Only weeding prevents them out-competing the aliens. If you abandoned the garden, it would slowly revert to a wild state, but would remain very unstable for years, its species composition constantly fluctuating, as delicate equilibria develop. A similar process would take place in a newly made pond if it was not freqeuntly cleaned out.

Ecological succession is said to take place when the vegetation (and associated fauna and micro-organism population) in a particular place changes with time, and one unstable community progressively gives way to another until a stable climax community becomes established. Often, the intermediate communities modify the environment in such a way as to create the conditions necessary for the establishment of the next, and, ultimately, of the climax (9.2 and 9.3).

A climatic climax community is a stable one which usually becomes established through a successional process. Its nature is determined by the climate and geology of the place. Chapters 3, 4 and 10 discuss important abiotic (non-biological) ecological factors as they apply on land, in water, and on the seashore. A climax community such as a natural (or semi-natural) forest represents a stable ecosystem in which thousands of plant, animal and microbial species interact, come into existence and die every day, and yet the whole thing represents a dynamic equilibrium which stays the same for thousands of years. Climax communities usually have a high species diversity (5.6), reflecting the numerous links of food chains, recycling and population controls which contribute to their stability.

The natural climax communities of over 60 per cent of Britain are considered to be various types of forest because of the temperate, oceanic climate. It is believed that this is the way our pioneering Stone Age ancestors found it and that it would revert to the same state if human influences were removed. Every time you neglect a garden, the long process in which the local climax reasserts itself begins. Trees will not grow above 500 metres in Britain, and here, the natural climax is grassland, bog or heath (8.4). Below the tree-line, climax is also affected by soil (3.2), as well as climate. Ash and Beechwoods are often the climax of limestone soils, Oak of the deep soils of the Midlands and Scots Pine of the upper parts of Scottish glens.

A dominant species is one with an over-riding influence determining the character of a community and ecosystem. Often it is the most abundant large plant species, and even gives its name to the community. Beech is dominant in a Beechwood, for example, where it produces the micro-climate (3.3), is the main source of primary productivity which supports the heterotrophes, and provides a dense shade which eliminates all but very shade-tolerant herbs. Dominants of heath are species of Heather (such as *Calluna vulgaris*), of pasture, the main grass species and in a peat-bog, the Bog Moss (*Sphagnum* spp.). If more than one species are equally dominant, they are described as 'codominant'.

1.7 THE HUMAN INFLUENCE IN BRITISH VEGETATION

The patchwork of fields and hedgerows of Lowland Britain was created and is still maintained by Man. Sometime before 3000 BC, Britain's vast primaeval forests began to be cleared for agriculture, fuel and building materials. This continued until by the nineteenth century little, if any, primary woodland remained, even in remote Highland glens. Most, if not all, present-day woodland has been planted, or, where it has regenerated naturally, has in some way been influenced by Man. Most grassland below 500 metres is maintained by grazing and mowing, and scrub usually implies successional changes back to woodland (9.2 and 9.3) following abandonment of pasture, railway lines or buildings, a reduction of rabbit populations, or destruction of tree populations by fire, felling or Dutch Elm Disease.

Agriculture has always involved interference with natural habitats, but at least in the past its changes were gradual, and ecological systems estab-lished some kind of equilibrium with it. Ancient hedgerows, pasture and hay-meadows often have developed as distinctive communities with a high species diversity, including numerous wild plant and animal species. Modern farming's economic efficiency and mechanisation involves removal of hedgerows, more intensive and frequent mowing, and the use of pesti-cides, all of which involve loss of species diversity.

There are few, if any, places in Britain where human influences, past and present, have not profoundly influenced the ecology. There is probably no truly 'natural' vegetation, but some, such as seashores, rough grassland, old ponds and woodlands, may be fairly stable climax or plagio-climax communities, and being regarded as 'semi-natural' are particularly suitable for ecological study. Historical information and details of current manage-ment should be regarded as important ecological data.

8

1.8 FIELD STUDIES AND ECOLOGY

This book does not deal fully with aspects of ecology which do not easily lend themselves to a sixth-form field course, such as energy flow and population changes with time. For these, consult other books, like *Ecology* by T. J. King (published by Thomas Nelson), and use computer simulations and laboratory-based experiments.

1.9 CONSERVATION WARNING

Collecting and trampling by ecologists can damage most sites, and NONE of the exercises suggested in this book need to be carried out in places with rare or sensitive species. Rare plants can even be damaged by the habitat-disturbance associated with careful botanical enthusiasts merely photographing them. Birds may desert their eggs long enough for the embryos to die just because you have been working near their nest.

(a) Keep collecting to the minimum needed for identification and avoid *any* species which you know or suspect to be rare. If in any doubt, consult your teacher or lecturer.

(b) Avoid working in sites which are heavily used by other groups of students. Some sites are more sensitive than others, and so, again, if in doubt, consult your teacher or lecturer.

(c) Never work on a site without getting the owner's permission. Not only is this normal courtesy, but he or she may point out things you need to know, e.g. possible threats to crops, lifestock, amenity, game or wildlife. Many landowners have Sites of Special Scientific Interest (SSSI's) on their land and share responsibility for protecting them with the Nature Conservancy Council (NCC) through Management Agreements.

(d) Remember that most wild plants, birds and many other species, e.g. badgers and bats, are protected by law.

2 Collecting Data

2.1 SAMPLING

If you wanted to find the number of daisies in a field, it would take a very long time to count them. It may take so long, in fact, that by the time you had finished, some of the first ones you had counted would have died. In scientific work, as in most other matters, time costs money, and, as the rain-soaked hours become rain-soaked days, you might well ask whether there might not be an easier way. The solution would be to estimate the size of the population of daisies in such a way that is sufficiently accurate for practical purposes but which involves the minimum of time spent in the field actually counting plants. If one could accurately estimate the number of daisies per unit area (that is, determine their density) and knew the area of the whole field, one could calculate an estimate of the total number of plants quite easily.

It may not be possible to examine every square metre of the field but selected parts of it could be looked at very carefully. A sample area, called a 'quadrat', could be marked out by placing a quadrat frame on the ground. If a 1 × 1 m quadrat were found to contain, for example, 10 plants, the density would be estimated as 10 individuals per m^2. Should the field have an area of 10 000 m^2, could it be claimed that it had 10 × 10 000 = 100 000 daisy plants? It probably could not, but it would depend on how evenly they were distributed and whether 1 × 1 m^2 happened to be the appropriate quadrat size. If the daisies were not in a field but formed a pattern on wallpaper, a single sample may be all that is needed, since the paper would have been printed by a machine which repeats the pattern at regular and frequent intervals, particularly if the quadrat corresponded in size and position to the repeating unit. In reality, dealing with real daisies in a real field, there is no reason to believe (and every reason to doubt) that the distribution of plants in the field is regular. There could be only 10 and if they were all in your quadrat, your estimated density of 10

10

plants per m^2 leading to an estimated population size of 100 000 individuals would be disastrously wrong. Clearly, your estimate of density should have been based on a considerable number of quadrats distributed throughout the field. Even so, counting daisies in 100 1×1 m quadrats is much easier than in 10 000. If the sampling is done corectly, your estimate could be as biologically useful as a complete count. But how do you decide where to put your quadrats?

Your first idea might be to decide (using skill and judgement you have yet to acquire) that a certain place 'looks pretty representative', in which case you are being subjective. Many people have a very unjustified faith in the validity of their subjective judgement. Subjectivity does have a place in carrying out surveys of plants, animals, soils, rocks and many other things in the hands of experienced operators, but they are very aware of the problems and use it with caution. The alternative is to seek to be objective, which means free of personal bias, meeting an important criterion of scientific work, namely that findings should be reproducible. With an objective method, any two people using the same method correctly should get a similar result. Quadrats should be located objectively, and there must be enough of them to provide a representative sample. There may be more daisies in the middle of the field than at the edge near the hedge, or near the public footpath across one corner, but, at the outset you cannot know for certain. No such variation should be assumed even if you (subjectively, of course) think it is obvious.

Quadrats are samples of vegetation or populations of more or less sedentary seashore animals. Soil samples are often collected in plastic bags and taken to a laboratory. Physical and chemical factors, such as light intensity and wind-speed are sampled by taking readings using various types of equipment (chapters 3 and 4) at certain places. Political opinion polls and marketing research involve selecting samples of people to be interviewed. Estimations of motile animal populations are more complicated than those for daisies (chapter 6), but these too require sampling. Using quadrats to estimate the density of common plants provides a useful introduction to the general principles of the subject. When sampling objectively, certain rules are fixed at the outset and are then followed slavishly even when they seem to lead to the recording of apparently foolish or irrelevant information. Quadrats which include cowpats must be recorded unless a decision was made to exclude them initially. You must not change the rules once you have started. If you decide the rules need changing, you must start again. There are two approaches to sampling, random and regular.

2.2 RANDOM SAMPLING

When placing 1 × 1 m quadrats randomly, each m^2 has an equal chance of being selected, even an equal chance of being selected twice.

(a) Throwing quadrats
This rather picturesque method has a long tradition in ecology. The quadrat frame is thrown and recorded wherever it lands unless this is outside the study area (such as over the hedge in the next field). The approach is not very objective because usually some parts of the field get under-recorded (for example, the corners). It can be improved by walking in a straight line and throwing the quadrat frame every so many paces (say every 10). The only advantage is that it is quickly done.

(b) Random sampling using a grid
A grid is a series of randomly spaced parallel lines with a second set of similar lines crossing it at right angles dividing an area into squares of equal size.

1. Decide where to lay out your grid. This (subjective) decision depends on the site and the objective of the study.
2. Lay out two 30 m tapes at right angles along two sides of the grid (a 30 × 30 m grid is a good size to practise the method, but you may, of course, modify this).
3. Regard each tape as a series of numbered metre intervals (1-30). Each of the 30 × 30 squares thus has two coordinates, one from each tape (figure 2.1a). Call one tape 'A' and the other 'B'.
4. Use a pair of random coordinates (see below) to locate your quadrat, and record it. Ideally, you should measure the position accurately at right angles to the tapes but in an exercise, using a grid as small as this, it may be adequate to judge the position by looking at both marker tapes. In a large-scale study, you should use surveying methods to mark the grid with parallel rows of pegs. Repeat until you have enough samples.

There are three ways you could obtain random coordinates.

1. 'Picking numbers out of a hat' — but instead of a hat, use a small (non-transparent) bag or box. If both sides of the grid are 30 m long, number 30 small, identical pieces of wood, large beads (or something similar) from 1 to 30. Place them in the bag and shake it. Remove one number

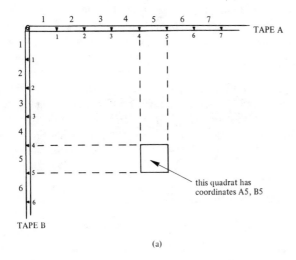

(a)

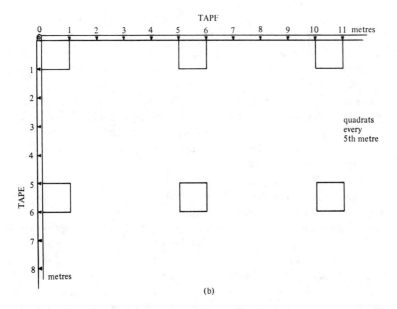

(b)

Figure 2.1 *Sampling using a grid: (a) Random coordinates, (b) Regular sampling*

13

as the coordinate on tape A. Replace it in bag, shake, and withdraw another (B coordinate). Always replace numbers before starting again.
2. Random coordinate tables (appendix A). Start anywhere in a column (if several students are sampling the same grid, they should all start in a different part of the table) and then take the coordinates in series. If one coordinate lies outside the grid (for example, if yours has sides of 30 m and a coordinate is 45) ignore the pair and move on to the next.
3. Use a computer. Computers (and some pocket calculators) can generate random numbers. Again, just ignore numbers which are too big for your grid. Alternatively, you may like to write a simple program to provide a series of coordinates in the required range. It is convenient to output the numbers through a printer and to take the printout into the field.

2.3 REGULAR SAMPLING USING A GRID: MAKING MAPS

Lay out a grid as described above (2.2). Instead of using random coordinates to place your quadrats, place them at regular intervals; for example, record every fifth metre along parallel rows 5 metres apart (figure 2.1b). A transect (8.5) is another form of regular sampling.

This method is easier to use, especially with a large grid, but, although it is equally as objective as the previous method, it has the disadvantage that there may be a regular pattern in the population being sampled which may 'resonate' with the sampling. To take an extreme example, cabbages in a garden may have been planted 50 cm apart. A grid of 0.25 × 0.25 m quadrats placed every 0.5 m, starting between the first two cabbages, could have the consequence that no quadrats are found to contain cabbages, giving a quite incorrect estimate of cabbage density of zero. Trees in a wood are usually planted in rows and meadows often have a mediaeval ridge and furrow pattern (9.8(b)).

Another disadvantage is that some statistical tests cannot be used with regular sampling. The Mann–Whitney test (2.6), for example, assumes that the samples have been collected randomly.

An advantage of regular grid sampling is that it can be used to draw a map, as illustrated in figure 2.2. Edmondston's Chickweed is endemic to the Shetland Island of Unst. This means it occurs nowhere else and is thought to have evolved there (probably from the Arctic Mouse-ear) in response to unusual local ecological conditions. It is, therefore, a rare species and so a conservation priority. A team of sixth-formers established a grid by putting permanent marker pegs 50 m apart and by determining the density of the species in each 50 × 50 m square. In figure 2.2 (a 'dot map'), there is a dot whose size is related to density corresponding to each

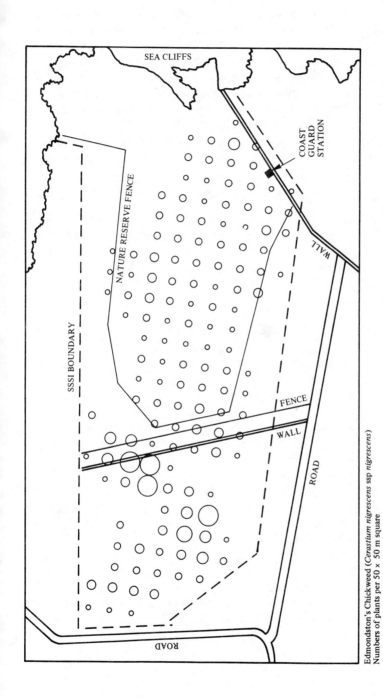

Edmondston's Chickweed (*Cerastium nigrescens* ssp *nigrescens*)
Numbers of plants per 50 × 50 m square

○ 1 – 10   ○ 11 – 50   ○ 51 – 100   ◯ 101 – 200   ◯ 201 – 300   ◯ > 300

Figure 2.2 *Dot map to display density of a single species within a gridded site*

15

square. The survey revealed that over half the World population of the plant grew outside the National Nature Reserve established to protect it. As a result, the Nature Conservancy Council enlarged the reserve. Because the grid was permanently marked it was possible to repeat the counting in later years and thus monitor changes. Numerous other species and soil chemical factors were also recorded using the same grid which, in addition, helped map geographical features such as the nature reserve boundary.

To make maps like this yourself, use the approach illustrated for iso-nome studies (figure 8.7). Note that if you are just making a map, you could use quite widely spaced quadrats, depending on local circumstances.

Exercise 2.1. Sampling density of common plant species

The object of this exercise is to familiarise yourself with sampling techniques. Biologically, the exercise should be kept simple, so that the whole class can cooperate in obtaining a large amount of very accurate data. Only very common, easily identified species should be recorded.

What you need
(a) Two 30 m tapes (could be shared).
(b) Quadrat frames of various sizes (1 x 1, 0.5 x 0.5, 0.25 x 0.25, 0.1 x 0.1 m).
(c) Record sheets.
(d) Random numbers (see 2.2).

Method
(1) Select two adjacent sites. A lawn and a meadow would be ideal.
(2) The class must agree on a few species to record. Daisies, Dandelions and Buttercups may suffice. Even such well-known species present problems and you must iron these out before you start. Dandelion (*Taraxacum officinale*) is not usually hairy, but some lawns may have plant species with similar leaves like the hairy Rough Hawkbit (*Leontedon hispidus*). There are three common species of buttercup found in lawns and meadows, Field Buttercup (*Ranunculus acris*), Bulbous Buttercup (*R. bulbosus*) and the Creeping Buttercup (*R. repens*), which you could record separately. You will perhaps need your teacher's help. No sampling, however statistically precise, has much value if the data are unreliable, and where a team is involved, every member must operate to the same standard.
(3) Lay out a grid as described in (2.2) on the lawn, and record the density of each selected species using different sizes of quadrat (10 of each size). Repeat the procedure in the other site (for example,

16

meadow). Several pairs of students could share a grid (there may be a shortage of tapes). Where there are several grids in the one site, arrange them to form a block (say three 30 x 30 m grids to form a combined 90 x 30 m one). To find density, simply count the number of individual plants in your quadrat. Even this is not completely straight-forward. Daisies, for example, usually grow in groups, and everyone must try to count separate rosettes (single emergent shoots).

In your choice of species: (a) restrict the number to five species, (b) use species that everyone can identify easily without examining flowers (that is, by vegetative characteristics) and (c) use species where you can count individual plants (for example, do not use grass).

Handling the data
There are three ways in which they could be considered

(i) effect of quadrat size
(ii) how large a sample was necessary (that is, how many quadrats?)
(iii) what do the data show about plant species in relation to the two sites?

Note that the word 'data' is plural. The singular does not exist in English — you have to refer to 'a piece of data'.

(i) Effect of quadrat size
Collate the class data (use blackboard) and calculate, for each site, the mean density for each species for each quadrat size

$$\text{Total no. plants/Total area of quadrats} = \text{Density}$$

For example, suppose that the class records a total of 100 quadrats on a lawn. These represent 100 m^2. These quadrats contain a total of 2622 daisy plants. The estimate of density is

$$2622/100 = 26.22 \text{ individuals per m}^2$$

If the combined gridded area was 30 x 60 = 1800 m^2, then estimated number of daisies in it = 1800 x 26.22 = 47 196 plants.
  Using the 0.5 x 0.5 m quadrat, suppose 100 quadrats contained 650 plants. Total area = 0.5 x 0.5 x 100 = 25 m^2. This gives an estimated density of 650/25 = 26.0, and a total population of the grid of 1800 x 26 = 46 800.
  Table 2.1 shows an example of the process.

Table 2.1 How to collect the class data (with some sample data)

Species: Daisy (*Bellis perennis*)

Habitat	Lawn				Meadow			
Total area of grid	1800 m²				1800 m²			
Quad. size (side in m)	1.0	0.5	0.25	0.10	1.0	0.5	0.25	0.10
Total no. plants	2622	650						
Total no. quadrats	100	100						
Combined area of quadrats (m²)	100	25						
Est. density (plants/m²)	26.22	26.00						
Estimated population	47196	48000						

In your write up

1. How does quadrat size affect estimates of density? Are some species more affected than others? If so, why? The answer may lie in the distribution or the size of the individual plants.
2. Were there any problems encountered in the collection of the data which might have reduced their reliability? Were there difficulties in plant identification or in deciding the number of individuals where they occur in groups?

(ii) Effect of sample size on estimates
Take one of the species (say Daisy), one quadrat size (say 0.5 × 0.5) and one of the sites (say the lawn), and use the whole class's data to calculate a series of cumulative mean values and plot them against sample size, as shown in figure 2.3. Take the first two quadrats and calculate the mean density at a sample size of two. Repeat this, each time adding an extra quadrat until you have included the data of the whole class. Table 2.2 shows you how to organise the process. Continue the table to include all

18

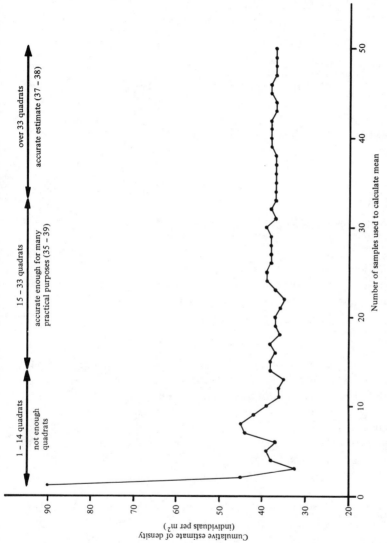

Figure 2.3 *Effect of sample size on the estimate of the mean. These data refer to a particular example concerning a particular species in a particular site. Do not expect your data to produce quite the same pattern*

19

Table 2.2   Calculation of cumulative means (with some specimen data)

Species: Daisy (*Bellis perennis*)
Quadrat size: 0.5 × 0.5 m (0.25 m²)
Site: Youth Hostel Lawn

Quadrat No.	Plants per quadrat	Density (per m²)	Cumulative mean density (Estimate of density)	
1	6	24		
2	10	40	$(24 + 40)/2$	$= 32$
3	1	4	$(24 + 40 + 4)/3$	$= 23$
4	25	100	$(24 + 40 + 4 + 100)/4$	$= 42$
5	15	60	$(24 + 40 + 4 + 100 + 60)/5$	$= 46$

the class data (50 quadrats in the example). It will be simpler to omit the calculation from your own table.

You should repeat this operation with the same species as recorded from the other site (separate graph), and then each of the other species using this quadrat size. If you have time, consider the other quadrat sizes. Convenient class organisation would be for each student to handle data for a particular species at a particular quadrat size. You could perhaps write a simple computer program to calculate cumulative means.

In your write up

Present the graphs you have drawn personally, but refer to your colleagues' conclusions as well. Figure 2.3 shows that at small sample sizes, the estimate fluctuates wildly with the addition of successive quadrats, indicating that sample size is too small to provide a reliable estimate. Above a certain number of quadrats, however, inclusion of more data has little effect on the estimate. In the example (figure 2.3), it appears that for the particular species, site and quadrat size the minimum sample size needed for a reliable estimate is approximately 33.

1. What was the minimum sample size needed for each of the species in your survey for each site?
2. Suggest reasons why some species require more quadrats than others, and, for a given species, why a larger sample may be required in one site than in the other. An explanation may include the size of the individual plants, their distribution and their overall density.
3. Which was the most suitable quadrat size tested?

20

(iii) Interpreting your data ecologically
Before dealing with this, read (2.4), (2.5) and (2.6).

In your write up

1. Use the combined class results from a particular quadrat size (the one which you decided was the most suitable in part (ii) of this exercise). For each species, calculate the mean density for each habitat; for example, in the example (table 2.3) the lawn had a mean density value for daisies of 36 individuals per m^2, whilst the corresponding mean for the meadow was 18.4. Find out if the differences in density are statistically significant (2.5). Present the results as shown in table 2.2.

Table 2.3   Density of five species in adjacent habitats

	Mean density (plants per m^2)		
	Lawn	Meadow	
Daisy (*Bellis perennis*)	36.0	18.6	S**
Dandelion (*Taraxacum officinale*)	10.6	3.6	S**
Field Buttercup (*R. acris*)	1.6	10.0	S**
Ribwort Plantain (*P. lanceolata*)	4.3	2.8	NS
Greater Plantain (*P. major*)	0.9	1.5	NS

S** = Highly significant ($p = 0.01$).
NS = Not significant.

   Summarise your conclusions in this style: "Density of Daisies and Dandelions was significantly higher in the lawn, but that of Field Buttercups was significantly lower. There was no significant difference in density of Ribwort and Greater Plantain."
2. Mention any other species which you noticed in your survey that you suspect may occur in different densities when comparing the two sites, and which might have been appropriate for inclusion in the study. Give your reasons.
3. Suggest hypotheses to explain differences in density of the study species between the two sites. It will be easier to do this after you have completed more of the Field Course and have gained some experience.

2.4 TESTING THE DIFFERENCES BETWEEN DATASETS FOR STATISTICAL SIGNIFICANCE

Does the data in table 2.4 permit the conclusion that the density of daisies in the lawn is more than in the meadow? The first step is to calculate the mean or 'average' (more correctly called the 'arithmetic mean') for both datasets. As you no doubt already know, the arithmetic mean is calculated by adding up the dataset and dividing by the number of values. It can be written as

$$\bar{x} = \frac{\Sigma x}{n}$$

where n = number of values in the dataset (10 in the example)
$\bar{x}$ = the arithmetic mean
Σ = 'the sum of'
x = each one of the values in the dataset taken in turn (in the example, x is a series of 10 values, one for each quadrat).

(a) Statistical significance

Strictly speaking, the mean values apply to the sample data, but are being used to estimate the mean density of the whole area sampled. You may have already found (part (ii) of exercise 2.1) that ten quadrats were not enough and so you should use the combined class data, and obtain more satisfactory estimates. There is undoubtedly a difference between the two means (one for each habitat) but are we entitled to attribute this to some ecological factor such as the fact that the lawn is closely mown with a lawn-mower weekly but the field is mown by a tractor with a large cutting attachment annually, or that the meadow (but not the lawn) is grazed by cows? We must first consider that the difference could be simply due to chance.

If all the quadrats in the lawn had the same density of daisies (all 50) and similarly, all the quadrats from the meadow were the same (18), we would have little doubt that the difference could not be due to chance, because there is no variation in each dataset. A glance at table 2.4 shows that in the example, this is not the case. In the lawn, the values vary about the mean with a range of 24 to 100 and those for the meadow between 4 and 32. The ranges overlap, but only to a limited extent. How much can the ranges overlap before we consider both datasets samples of the same population – that is, before we regard any differences between the means as having no statistical significance?

22

Table 2.4 Comparing means and datasets

Species: Daisy (*Bellis perennis*)
Quadrat size: 0.5 × 0.5 m (0.25 m^2)

Lawn		Meadow		Rank values for Mann–Whitney test*	
Plants/ quadrat	Density (per m^2)	Plants/ quadrat	Density (per m^2)	Lawn	Meadow
6	24	4	16	8.5	5
10	40	5	20	14	6.5
11	44	3	12	16	3.5
25	100	5	20	20	6.5
15	60	1	4	18	1
13	52	2	8	17	2
19	76	3	12	*2¹⁹*	3.5
10	40	8	32	14	12
6	24	7	28	8.5	10.5
10	40	7	28	14	10.5
Total (Σx)	500		180	132 *149*	61 (*T*1)
Mean ($\bar{x}$)	50.0		18.0		

*In the Mann–Whitney test (2.6) *T*1 (the value you look up in the table) is the sum of the ranks of the dataset with the lowest total rank value. In this case, the sum of the ranks for the meadow dataset (61) is the lower of the two and so *T*1 = 61.

The variation within the datasets is partly caused by chance. If we were to repeat the sampling, it is quite possible none of the random quadrats would occupy any of the positions used in the previous survey (although if the sample were big enough we would get a similar mean). The larger the variation within the two datasets, the bigger the difference between the means must be to be statistically significant. The word 'significant' has a very specialised meaning in statistics. If the difference between means (or a correlation coefficient, 8.9, or a chi-squared value, 5.7) is significant, it means we assume it cannot be explained by chance alone.

(b) The null hypothesis
In the logic of statistical tests, all differences are assumed to be 'not significant' unless they can be shown to be otherwise, as, in British Law, you are innocent until proved guilty. In England and Wales, if you cannot be proved guilty, you are acquitted, without any blemish on your character, even if this was due simply to lack of evidence. Similarly, a 'verdict' of

23

'not significant' may leave suspicions in your mind but you must seek more evidence (more data) before you can proceed. In Scottish Law, there can be a third possibility, a verdict of 'not proven', and this is a better analogy for 'not significant'.

Where chance is concerned, absolute certainty does not exist, but statistics attempts to quantify certainty by expressing probability. Whether two sets of data are significantly different can be tested. The test takes into account the difference between the datasets and the variation within the datasets used to calculate them. To be significant, a difference must be so big that the probability of its being explained by chance is less than 5 per cent – that is, above the 5 per cent significance level (or $p = 0.05$). 'Highly significant' and 'very highly significant' signify that the probability of the difference being explained by chance is even less ($p = 0.1$ and $p = 0.01$ respectively). If a result is shown to be significant, the null hypothesis is rejected, and only then can we seek another hypothesis. Only when we have rejected the null hypothesis can we start to explain the differences in the density of daisies in the comparison of lawn and meadow and only then can we start speculating about mowing, grazing, soil moisture and nutrients.

2.5 TESTS OF SIGNIFICANCE: PARAMETRIC VERSUS NON-PARAMETRIC

Many common statistical tests that you may have already met in mathematics courses are described as 'parametric' and are unsuitable for field data because they make assumptions such as that the data are normally distributed. Field data often contain numerous zeros, and are often non-normal. What 'normal' means need not be discussed here, but you should appreciate that the tests in this book have been selected for their suitability in the field (and are 'non-parametric'). Treat others with caution. In particular, if using a computer program, you should consult your teacher to check that the tests are appropriate for your data.

2.6 THE MANN–WHITNEY TEST FOR COMPARING DATASETS

This is non-parametric and more suitable for field data than the commonly used (parametric) t test. It assumes that the data have been collected randomly. Strictly speaking, the Mann–Whitney test compares datasets rather than means, but this statistically nice distinction is unlikely to be a problem to you and you can regard it as a method of comparing your means.

1. It is suggested that when you decided how many samples to collect that you arrange things such that both datasets have the same number of values (for example, in table 2.4 there are 10 samples from the lawn and 10 from the meadow). The Mann–Whitney test can deal with comparison of different sized datasets but the process is slightly more complicated and you need a more elaborate table to interpret them than that in appendix B. To deal with different sized datasets, consult any statistical textbook which includes non-parametric methods. Also note that appendix B only deals with a limited number of dataset sizes between 50 and 100.
2. Arrange both datasets (combined) in order of rank. Mark the numbers so that you remember which of the datasets each value came from. This is similar to positioning examination results except that the smallest value has a rank of 1.

   Using table 2.4 data

Rank	1	2	3	4	5	6	7	8	9	10	11	12	13	14	15	16	17	18	19	20
Value	4*	8*	12*	12*	16*	20*	20*	24	24	28*	28*	32*	40	40	40	44	52	60	76	100

   where the asterisk indicates values in one dataset (in this case, meadow), and the unmarked ones are the other dataset (lawn).
3. Some values tie for their rank with others. For these calculate the 'mean rank'; for example, ranks 13, 14 and 15 all have 40. They all share a common mean rank: $(13+14+15)/3 = 14$. Similarly $(6+7)/2 = 13/2 = 6.5$. Large numbers of tied ranks reduce the sensitivity of the test.
4. Separate the rank values into two columns, one corresponding to each dataset as in table 2.4.
5. Calculate the total rank for each column (table 2.4).
6. The lowest of the two total rank values is $T1$. In table 2.4 $T1 = 61$.
7. Consult the tables of significance levels (appendix B). The 5 per cent column gives a tabled value of $T1$, with 10 pieces of data, of 78. As our calculated $T1$ value is less than this, the null hypothesis is rejected and the difference between the datasets declared significant 'at the 5 per cent level'. The 1 per cent column shows that the difference is even more significant because the calculated $T1$ (61) is less than the tabled value of 70. The difference between the datasets is significant 'at the 1 per cent level' or 'highly significant'. Had it been, for example, 77, it

25

would have 'scraped home' at the 5 per cent level ('significant', $p < 0.05$) but not at the 1 per cent level (not 'highly significant', $p < 0.01$).

What this means is that, although you can never be absolutely certain that the difference is not due to chance, the probability is sufficiently small for it to be considered negligible. According to appendix B, the probability that the difference can be explained by chance is less than 1 per cent. The probability of $T1$ being as low as 70 as a result of chance alone is only 1 per cent, and our value is even lower. Now that we have disposed of the null hypothesis we can start looking for a biological reason to explain the difference in the density of daisies in the two places suggested by our data.

2.7 SAMPLING: SOME BASIC PRINCIPLES SUMMARISED

Quadrats provide a very good introduction to the principles of making estimates based on sampling in other fields. For example, to estimate the density of red blood cells in a blood sample, a drop is placed on a haemo-cytometer, a microscope slide with a grid etched on it.

1. Define the limits of the population
If you are sampling a field, it can be defined as all the land enclosed by the hedge, wall or fence. When you lay out tapes to mark out a grid, you also precisely define a study area. If you are sampling randomly, every part of the designated area has an equal chance of being selected.

2. Sample as objectively as possible
Some degree of subjectivity usually creeps in somewhere. This may be unavoidable, but you should be aware of it. In exercise 2.1, your choice of where to put the grid and which species to record were subjective. Plant identification usually involves some subjectivity, as does deciding how many individuals there are in a dense clump of daisies. Objective methods are often time-consuming and recording schemes often include a subjective element (for example, in estimating cover of plant species (5.3). The aim should be to minimise subjectivity, but a compromise between objectivity and practicality is usually unavoidable.

3. Base estimates on a big enough sample
There are ways of determining minimum sample size based on a preliminary study (as illustrated in exercise 2.1). In practice, however, practicalities of time, equipment and expense often lead to sample size being another com-

promise. Exercise 2.1 was intended to develop your sensitivity to the problem, and, if in doubt, aim for as big a sample as possible. Note, however, that an unnecessarily large sample represents a waste of time which might be better employed in some other way.

4. Collect data accurately
Sort out techniques and identification before you start.

5. Be consistent
Once you have agreed on a procedure, stick to it.

2.8 RECORDING SHEETS AND SPECIES LISTS

Many of the exercises in this book include an example of a recording sheet. The format of a well-prepared record sheet helps you to organise your data as you collect them and facilitates interpreting them later. If the data are being collected by a team (such as when your class works together), design of the sheet assists collation too, and preparing it together ensures that you make certain joint decisions in advance. An example of this is making a species list. You decide which species to record and everyone has them on their sheets in the same order. The notes which follow table 8.1 may be helpful in this. In preparing a sheet, consider what you intend to do with it later, including any statistical procedures that you may wish to apply.

When deciding which species to record, you must be realistic. When working in a new site, even an experienced ecologist has difficulty in knowing all the species, especially by vegetative characteristics (that is, by leaves and stems) when plants are not in flower. When people work as a team, some are more accurate in their identifications than others, and the combined data may be only as good as the most incompetent recorder. Do not attempt to record all species. Ignore those that are met very infrequently (unless you have reason to think they are important), and avoid pretending to distinguish ones which you know you cannot. In the example (9.8) there were specific objectives in mind, and only three species were recorded. See also exercise 2.1 (method point (2)) and notes following table 8.1. You will need your teacher's or lecturer's help.

You will find a clip-board useful for record sheets, instructions and identification notes. Figure 2.4 shows the 'all-weather' clip-board which is enclosed by a large plastic bag for use in the rain.

27

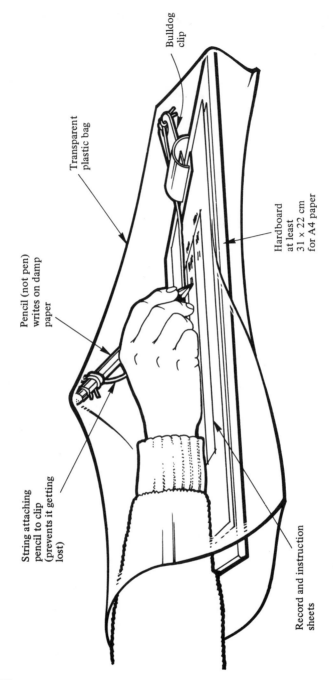

Bulldog clip

Transparent plastic bag

Hardboard at least 31 x 22 cm for A4 paper

Pencil (not pen) writes on damp paper

String attaching pencil to clip (prevents it getting lost)

Record and instruction sheets

Figure 2.4 *The all-weather clip-board (drawing based on a design by Nicola Probert)*

2.9 MICROCOMPUTERS IN HANDLING ECOLOGICAL FIELD DATA

Computers transformed ecology by making it possible to handle the large volumes of data required for a rigorously objective, statistical approach. Microcomputers can be very useful for Field Work, especially if you take one or more with you on your course, to use at your study centre. Remember to use the computer as a slave not a master. Use it to help you do whatever you would do anyway, but to do it quicker and on a bigger scale.

You (or another member of your class) may be able to write programs ('software') especially for your field course and several suggestions have already been made concerning this. Mostly, however, you will use commercial programs, such as 'Ecosoft' (for the BBC Model B). The authors can supply 'ecostat' for the Sinclair Spectrum which is particularly suitable for the exercises in this book.

(a) Entering the data
This may take longer than you expect. If often helps if one person reads out the numbers whilst somebody else enters them into the computer. You will make mistakes, and entering a whole class's transect data at once can be a frustrating business. Some programs have facilities for extending existing files. The authors' 'ecostat' program has an accompanying one which does this. Each pair of students can enter and file their own data, and then all the files can be combined to consider the class data as a whole.

(b) Filing
When you switch off a computer, it 'forgets' all the data stored in its memory. 'Filing' involves recording it on tape cassette or floppy disc (or Sinclair microdrive) separately from the program for further consideration on a future occasion. Once you have entered your data, get them safely on file before proceeding.

(c) Sorting and displaying data
A transect can be represented by drawing histograms (figure 8.3) or kite diagrams (figure 10.2) and an isonome study or simple map using the 'pen and paper' method of figure 8.7. Scatter diagrams can be plotted by hand and 'best fit' lines added subjectively (figures 8.4 and 8.5). A computer can do all these for you very quickly (a printer can provide 'hard-copy' on paper), and this means you start to discuss the ecological significance of your data soon after they have been collected. Some programs draw three-dimensional 'solid' diagrams which it is very difficult to draw by hand.

29

(d) Statistical tests

The same program which files and presents data will also carry out a variety of statistical tests which can be very tedious to do on paper, especially with the large volumes of data needed for good results. The computer can enable you to use these tests routinely and be a valuable 'slave' in getting the best out of them without your Biology field course turning into a mainly mathematical exercise. Avoid-using a computer for procedures that you do not understand. It is a good idea to practise any statistical procedures on paper on a small scale before using a computer to do them. Be careful to note which means of calculation your commercial program uses (see section 2.5). Programs involving rank ordering (such as for the Mann–Whitney test and for Spearman's correlation coefficient) may be rather slow, and with small datasets it may be quicker to do them with pen and paper.

2.10 MICROCOMPUTERS AND DATA COLLECTION

Mention is made in several parts of this book to the use of electronic devices (sensors) to measure such things as oxygen (4.2), light (3.5) and temperature (3.8). You may, for example, wish to measure oxygen levels in a pond over a 24 hour period. You could do this manually, but the tireless, unsleeping computer can do this for you. Connecting devices to a computer is called interfacing, but another piece of equipment (an interface) is required. Sensors give out a varying yet continuous voltage (analogue) output, but you require the data in a digital form (that is, as a series of separate readings, of, say, oxygen concentration every 15 minutes). An 'A to D' (Analogue to Digital) interface effects this conversion. Some computers such as the BBC Model B have a built-in A to D converter, with sockets at the back known as a 'user port'. The Sinclair Spectrum does not have this, but Griffin and George market an attachment which provides it with one. Philip Harris also supply various interfaces. You will require some kind of software to use with them (usually supplied with them or perhaps written by yourself). Computers can run on batteries, although it is not usually convenient to use them in the field but only for laboratory experiments (chapter 11). In the field, you need yet another piece of equipment (or 'hardware'), a datamemory.

The Philip Harris Datamemory* has its own in-built rechargeable batteries and can collect up to 512 readings from a variety of sensors over a period of up to one week. It has two channels (that is, it can make separate readings from two sensors at once) and so you could, for example, take

---

*This Datamemory has been superseded by the Philip Harris Easy Memory Unit (EMU) which has 4 channels and is interfacable. As its name implies, it is easy to use, especially in the field.

simultaneous recordings of dissolved oxygen and temperature in a pond over 24 hours. The device is not primarily intended for interfacing, but this can be done via a suitable interface (its playback output is digital), but you must write your own software. It is normally played back through a chart-recorder or voltmeter.

A 'VELA' is a four channel datamemory with built in A to D converter which interfaces through a BBC Model B user port (or separate interface for the Sinclair Spectrum). You can also playback through its own internal digital voltmeter or a chart-recorder. Unlike the similar Unilab device, it can operate from batteries in the field. Make sure your VELA is the type which has internal batteries to prevent it from 'forgetting' the data whilst in transit back to the laboratory.

The software (e.g. the Philip Harris DATADISC program, particularly suitable for the non-computer-minded wishing to use VELA) for such devices can display the stored data as graphs, and so your ever-versatile computer becomes a chart-recorder. When used in the laboratory, this avoids having the computer committed to one activity for extended periods. The main disadvantages are expense and vulnerability to vandalism. The devices can only be used on private land. They are not weather-proof but can be protected by covering them with a large, upturned fish tank, or, for short periods (such as overnight) by enclosing them in a large plastic bag.

If you are interested in electronics, you will find how to make your own sensors from magazines on the subject.

3 Measuring Environmental Factors on Land

3.1 LAND AS A HABITAT

It is believed that over 390 million years ago, plant species evolved which colonised land. The main advantage in this was probably the greater availability of the light necessary for photosynthesis. Animals quickly followed, exploiting the vast supply of food offered by this expansion of the World's vegetation. In many respects, however, land as a habitat was, compared with the sea, inhospitable. Water supply depended on rainfall, and was very variable, as was pH, mineral salt supply and temperature. In many parts of the World, temperatures sometimes fell below freezing point, and in others, rose above $45°C$, threatening to denature proteins. Buoyancy provides support in aquatic habitats, but on land not only is this support lacking, but windspeeds can reach over 160 km per hour.

3.2 SOIL

Soil is a complex feature of the terrestrial habitat, without which much evolution of land-plants could not have taken place. Soil, consisting of a mixture of rock fragments of various sizes and decaying organic matter, is essentially particulate. Plant roots can grow into it and provide anchorage for extensive aerial shoot and leaf systems. Water is held by soil and yet soil is at the same time porous, allowing the entry of oxygen. Soil particles also retain mineral ions, and provide a habitat for numerous micro-organisms and invertebrates which recycle plant nutrients. Soil offers at least some of the stability offered by ocean, and yet it varies much more from place to place. This is mainly because it is derived from rock, which varies greatly in chemical composition and physical properties, even within one small locality. Topography is a further complication, which affects drainage and surface stability.

3.3 MACROCLIMATE AND MICROCLIMATE

Macroclimate refers to the overall climate of a region. For example, the macroclimate of South West England is temperate and oceanic. Winter temperatures are mild, with less frost than in the Central Scottish Highlands, and the growing season is longer. Rainfall is plentiful, but droughts in July and August are not uncommon. Shetland, 160 km north of the Scottish mainland, being even more oceanic than mainland South West England and similarly influenced by the Gulf Stream, has less frost, but a shorter growing season, cooler summers and more wind.

Microclimate refers to local conditions which may vary over very short distances, even between opposite sides of a tree trunk, since only one side is exposed to the prevailing wind. In many parts of the World, the natural vegetation is forest. Even a small wood has a far more stable microclimate than an open field, since the trees offer shelter from wind, and humidity, temperature and water supply are more constant. This provides the organisms which live in the wood with some of the stability of the ocean habitat.

3.4 LIGHT AS AN ECOLOGICAL FACTOR

Ecosystems require light energy (1.4), and the amount of light available affects its primary productivity, and thus profoundly influences the rest of the system. Total solar radiation in the growing season and day-length are related to latitude but you are unlikely to be concerned with such variation on a field course. Within a particular area, the degree of shade, usually provided by the larger plants, but also by buildings, walls and cliffs (related to aspect), can cause considerable variation in ground flora, within a small area as sampled in a transect or isonome study (exercises 8.2 and 8.3).

3.5 LIGHT IN WOODLANDS

The dominant species have a profound effect. Pine trees, especially when closely planted as in a plantation, cast such a dense shade that the woodland floor is largely devoid of ground flora. Unlike the evergreen Pine, the deciduous Beech only casts a dense shade once leaves have formed. In spring, there is a 'light phase', when numerous species of herbs like Bluebells (*Hyacynthoides non-scriptus*) and Snowdrops (*Galanthus nivalis*) usually helped by organs of perennation like bulbs and corms) quickly complete their life-cycles. If you visit a wood in summer, look for fruits of such species, which may be all that remains to remind you of the seasonal dimension. Consider the whole wood as a photosynthesising unit. Beech absorbs most of the light itself, leaving little for the ground flora. Ash (*Fraxinus excelsior*) casts a less dense shade, and shares the task of absorbing light with a rich ground flora, and also a shrub layer (figure 3.1). As with the Beech and Pine, little light is 'wasted' by being allowed to reach the ground surface.

Exercise 3.1. Lightmeters in short-term studies

The Griffin Environmental Comparator is a robust battery-operated instrument to which light (and temperature, 3.8, 3.9), probes can be attached.

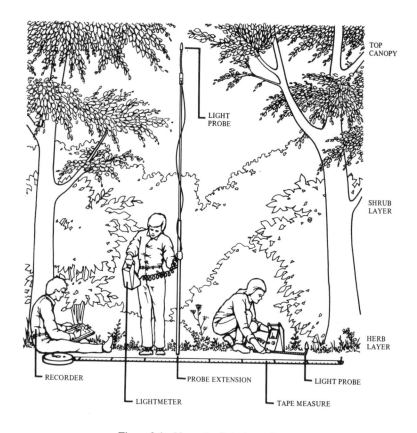

Figure 3.1 *Measuring light intensity*

The special "Field" light probe has a flexible tip ending in a small solar cell.

When taking readings, note the following points.

1. The solar cell must always be pointing in the same direction (usually vertically). This can be achieved when using it close to the ground by bending the tip (see figure 3.1).
2. Height above ground must be constant. Scrub and woodland usually represents a layered structure (figure 3.1 and section 3.5). In addition to taking readings just above the herb layer, at each sampling point, measurements could be made at various heights above ground, as far as you can reach, using the extension rods available for the probe. On a graph describing a transect, plot light intensity at each of the heights at which you have sampled as separate lines on the same graph. When comparing sites, calculate separate mean values for each height.
3. External light source must be constant throughout (3.6). There is no point in taking a series of measurements to compare light intensity between different places if the sun keeps disappearing behind a cloud every few minutes. In this kind of study, you need a cloudless day, and all readings must be made as quickly as possible.
4. Calibration. Unless a lightmeter calibrated in, for example, joules, lux, or foot-candles, can be borrowed, and the comparator calibrated (as with temperature calibration, exercise 3.4), take direct readings from the instrument, and refer to them as 'arbitrary units'.
5. Sensitivity. Adjust the sensitivity to give a full-scale deflection in the lightest part of the study area, and do not alter sensitivity during a series of measurements.

3.6 LONG-TERM STUDIES OF LIGHT INTENSITY

Using a lightmeter may be the best you can do, but you should appreciate its limitations in a constantly varying situation. There will be patches of bright light on the ground. These sunflecks move as the Sun moves during the day. This could make a few lightmeter readings made on one occasion valueless, although a large number would still give the general picture. It would be more satisfactory to measure light intensity over a longer period, preferably from dawn to dusk, or at least over several hours. Light sensors attached to a datamemory would overcome the problem, but you are unlikely to have enough equipment to make simultaneous measurements in more than a few places at once. Electrolytic integrators can do this cheaply. They are not commercially available, but the authors can provide instructions for their construction and use.

3.7 LIGHT COMPENSATION POINTS

The light compensation point of a particular plant species is the light intensity at which the rate of carbon dioxide production due to respiration equals the rate of carbon dioxide consumption production due to photosynthesis and at which, consequently, there is no detectable gas exchange. Since respiration involves the consumption of organic matter which must at some time have been produced by photosynthesis, the light compensation point can be regarded as a 'break-even' point. During the night, all plants are below compensation point, consuming stored organic material, usually starch. Every plant must have sufficient light to be above its compensation point sufficiently to meet its requirements for respiration and have organic matter to spare if growth is to be possible. In shaded habitats, such as woodlands, the low light intensities exclude many species, but shade species such as Dog's Mercury (*Mercurialis perennis*) are adapted for photosynthesis in such conditions. These adaptations give shade species a lower light compensation point than sun species found in more open habitats.

Exercise 3.2. Comparing light compensation points

In this study, one or more species are selected which seem to be restricted to woodland and one or more from open habitats near the wood, and the hypothesis is that they differ in light compensation point. The procedure is a familiar one of 'O' level Biology courses in which leaves are enclosed in glass specimen tubes with bicarbonate indicator, as illustrated in figure 3.2. Select species whose leaves are of a similar size, and large enough to wedge in the upper part of the tube. Best results are obtained on a sunny day.

What you need
(a) As many specimen tubes as possible (the longer the better) with plastic caps or rubber bungs.
(b) A supply of bicarbonate indiator (working strength).
(c) A supply of leaves all more or less mature, and of similar size. You should have at least one species from the wood, and one from the open habitat, but more than one from each place would be better. They should be freshly collected as you start to set up the experiment.

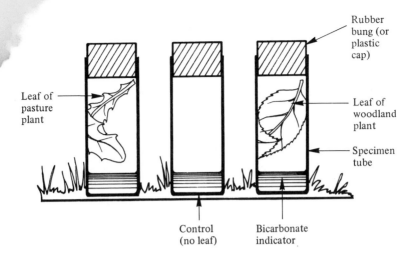

Leaf of pasture plant

Rubber bung (or plastic cap)

Leaf of woodland plant

Specimen tube

Control (no leaf)

Bicarbonate indicator

Figure 3.2 *Comparing light compensation points*

Method

Put about 1 cm depth of indicator in each tube. The solution is a mixture of pH indicators, and if it changes from red to yellow as soon as you put it in a tube, it means the tube is dirty. If this happens, rinse it out and add fresh indicator. Set up the tubes as in figure 3.2, and place a set of tubes (including one of each of the selected species) at measured intervals along the transect or on an isonome grid where there is a known light gradient (for example, pasture-scrub-wood). Leave for several hours, and observe the colour of the indicator. A control tube at each point (or one for the whole series) without a leaf, in which you expect no change, is used as a reference. Judge whether the indicator in the tubes with leaves is more yellow, or more purple, than the control.

Interpreting your results

At the darkest point of the transect, the indicator may turn from red to yellow in all the tubes, indicating that carbon dioxide production exceeds carbon dioxide absorption, and so all the species were below their light compensation point at that particular place on that particular day. At the lightest end, the indicator in all the tubes may have changed from red to purple, showing net carbon dioxide consumption, and all species above their compensation point. Since the transect has been selected to cover a range of light intensity, there should be a point somewhere along it where the indicator in the tubes is yellow in some tubes but red or purple in some of the others. The species which have caused the indicator to change to yellow have fared less well under the restricted light intensity and have a higher light compensation than the others. If the light intensity is a gradual one between extremes and you have sampled frequently, you may find, in fact, that each species has a different compensation point. It is assumed that you have done this as part of a study of vegetation and in conjunction with light measurements (as described above), in which case you will be able to relate your results to the actual distribution of the species in the field (chapter 8).

3.8 AIR TEMPERATURE AS AN ECOLOGICAL FACTOR

1. Metabolism depends on enzymes which have temperature optima. Individual species may be adapted for particular temperature ranges. The enzymes of homoiothermic animals are protected from environmental temperature extremes, but only at the expense of physiological stress leading, for example, to increased food requirement in winter.
2. Cells are damaged if their cytoplasm freezes. This is obviously a problem in winter, but night temperatures often fall to near freezing point in summer, especially in open habitats.
3. A rise in temperature tends to increase the rate of transpiration, aggravating any moisture stress suffered by plants (3.15).
4. The number of months of the year when the temperature is above a a certain critical level determines the length of the growing season, which affects the distribution of species. This is reduced by increasing altitude and latitude.

The main aspect of air temperature that you can readily study concerns the relationship between living organisms and microclimate. Air temperature varies rapidly, and the degree of change over a day is an important feature of microclimate. This is more important than the temperature at any one moment, and can only be studied by making a series of measurements at the same place (or places) at intervals. If possible, attach several probes to a datamemory, and leave for 24 hours, or better still, several days (2.10), with at least one probe in the wood, and one in the open. The data you obtain will put your other measurements in perspective.

Exercise 3.3. Measuring air temperature

Always avoid direct sunlight falling on the thermometer or end of the probe. Electronic probes have the advantage that they can be placed on extension rods and so measurements can be made at various heights above the ground, where you otherwise could not reach. The Griffin Environmental Comparator, fitted with a temperature probe, is particularly suitable. As with light measurements, air temperature is very variable, and so you must make your measurements on a day when there is not intermittent cloud cover, and make them quickly. As with light, take your readings at several, pre-decided, heights above the ground at each sampling point, and it is also important to decide the best sensitivity range before you start, and to stick to it throughout. When you present your results as graphs, convert your readings (arbitrary units) to degrees C by constructing a calibration curve (exercise 3.4). If you have time, repeat your measurements, and take them again as many times as possible over a day.

Exercise 3.4. Calibrating a temperature meter for environmental use

This exercise primarily concerns temperature, but the principles it illustrates apply to all forms of calibration. Each type of device has its own peculiarities, and it is suggested that you carry out a calibration procedure in the laboratory before going in the field, so that you familiarise yourself with the equipment.

What you need
(a) A thermometer (0° to 50° or 0° to 100°C range).
(b) Beakers containing water at a range of temperatures (use iced water to get a temperature between 0° and 5°C).
(c) A Griffin Environmental Comparator with temperature probe, or similar device.

Electrical temperature measuring devices usually
(i) do not give direct readings in degrees C but in arbitrary units or volts and
(ii) have some means of adjusting the sensitivity range.

Interpreting arbitrary units and giving them a real value (such as degrees C) is called 'calibration'. The instrument must be calibrated for the particular sensitivity range that you are using.

The general idea is to place the probe in several water samples whose temperature has been measured accurately using a thermometer. These samples, whose values have been accurately measured for calibration, are called 'standards'. The probe is placed in each standard in turn and when the meter reading becomes stable it is recorded together with the thermometer reading. Finally, the temperatures (measured by thermometer) are plotted (horizontal axis) against meter readings, and the 'best fit' line provides a calibration graph or calibration curve (figure 3.3). This graph can now be used to interpret field meter readings for which you have no corresponding values obtained with a thermometer.

You have to decide on the range you want to cover and the sensitivity setting. The range must include a standard below the minimum value (in this case temperature) that you expect to encounter in the field, and one above the expected maximum. A range from iced water to 55°C is adequate for warm weather in Britain, but often (unfortunately!) a lower maximum will be appropriate. How many standards you need depends on whether your sensor gives linear readings (that is, whether the calibration graph is a straight line) as does a Philip Harris temperature sensor, or a curvilinear one, as does a Griffin Environmental Comparator (figure 3.3). If the readings are linear, only two standards (top and bottom of the range) are needed, but three (one mid-range) would be safer. Non-linear readings need more standards between the extremes, (say five or six). If the sensor gives linear readings, adjust the meter to make your range use the whole scale, giving maximum sensitivity. With the Griffin instrument your temperature range should be, as far as possible, confined to the more

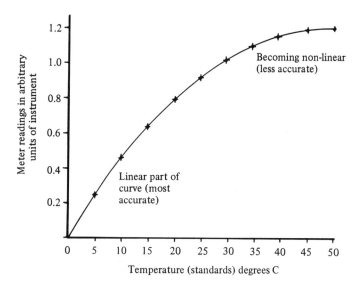

Figure 3.3 *Calibration graph for electronic temperature meter. The particular instrument used here has a non-linear output*

linear part of the curve (left-hand side), and you need to select the sensitivity setting (A to E) which gives about a half-scale deflection with your highest standard. When using the Griffin device, either select one sensitivity setting and stick to it, or use whichever gives a good deflection, and calibrate each sensitivity scale separately (making sure that you record the sensitivity setting used for each measurement).

3.9 SOIL TEMPERATURE AS AN ECOLOGICAL FACTOR

1. Soil temperature is ultimately related to air temperature, but it fluctuates less and so often gives a better indication of microclimate related to vegetation and altitude. Roots, seeds, bulbs, rhizomes, microbes and soil animals are protected from changes in air temperature by being in soil.
2. Soil temperature is often related to water content. Clay soils (3.2) which store more water than sandy ones, take longer to warm up, and this is particularly important in spring, when seed germination must await a critical temperature which differs between species. For similar reasons, drainage may affect soil temperature.
3. Aspect. South-facing slopes are warmer than north-facing ones.

3.10 MEASURING SOIL TEMPERATURE

A soil thermometer, illustrated in figure 3.4, has a metal casing which enables it to be inserted into the ground. It will, however, break if forced in. Measure soil temperature at the same points along a transect, or on an isonome grid, at two-hour intervals during a day. To compare readings made in this way, the thermometer must always be inserted to the same depth. Measure temperature at several depths at each sampling point. The depths you use depend on how easy it is to insert the thermometer in the particular soil being studied. Choose the depths after exploratory probing.

Exercise 3.5. Temperature and soil depth − a more detailed study

What you need
(a) A Griffin Environmental Comparator (or similar device) or field data-memory with several inputs.
(b) 2, 3 or 4 temperature probes.
(c) A spade or a trowel.

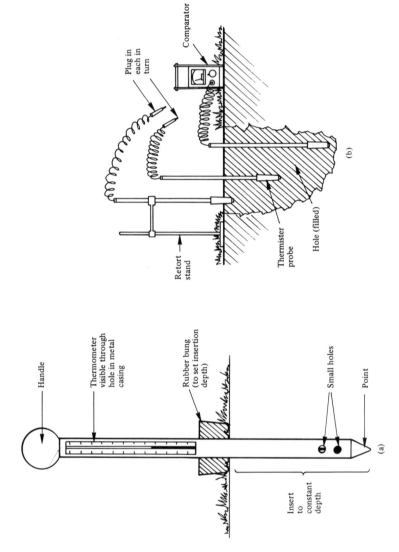

Figure 3.4 *Measuring soil temperature: (a) a soil thermometer, (b) making repeated measurements at different depths using thermister probes*

42

Method

Best results are obtained on a hot day in an open habitat. Figure 3.4 shows how to use an environmental comparator. Temperature probes cannot be driven into the ground. You must dig a pit (the deeper the better!) and refill it, burying the ends of the probes at measured depths — for example, 0 (surface), 20, 40 and 60 cm (depending on the depth of the hole) — leaving the wires emerging from the surface, and labelling each probe on its jack-plug. Set this up in the morning, and read the meter every 15 minutes for as much of the day as possible, plugging in each probe according to a definite sequence. Before taking recordings, wait until the needle has settled down and is giving a constant reading. Choose one sensitivity setting and keep to it throughout. Make sure that the surface probe is protected from direct sunlight (such as under leaves).

Before the data can be graphed (temperature against time with a separate line for each depth, all on the same graph), each probe must be calibrated (exercise 3.4).

Using a datamemory (2.10)

The procedure is similar to the one above, but you save the time in taking readings, and can take readings for 24 hours or longer.

In your write up

1. Consider temperature variation along the transect etc.
2. Consider temperature variation with depth at any one time, and over a day.

3.11 WIND AS AN ECOLOGICAL FACTOR

1. Mechanical damage: high winds may break or uproot large plants, and, as a result, some species may be absent from exposed places.
2. Desiccation: wind has a desiccating effect and hence increases transpiration rate. Plants in exposed places often show xeromorphic adaptations (3.15). This is true of many mountain species, such as Edelweiss, where the wind may be the main cause of moisture stress.
3. Near the sea: here the air is laden with salt spray. The salty solution carried by the wind accentuates the desiccating effect, by osmosis, giving rise to an effect resembling scorching amongst crop plants. Many maritime terrestrial plant species, such as the Sea Plantain (*Plantago maritima*), have xeromorphic features which enable them to survive in such conditions. Trees near the sea may show flag growth. Their shoot systems bend away from the prevailing wind, because the salt in the wind kills buds on the exposed side.

Exercise 3.6. Studying wind speed

What you need
(a) An anemometer (figure 3.5) or a windmeter (figure 3.6).
(b) A means of measuring up to 2 m.
(c) A pipe-cleaner (to clean and dry inside the windmeter).

Method
To use a windmeter, stand facing the wind and hold up the device with the 'front' facing towards you. It must be operated dry, and so cannot be used if it is raining. With both devices (anemometer and windmeter) it is easy to take readings when the wind speed is fairly steady, but gusting may be a problem. How you cope with this depends on the circumstances, and you must ask yourself whether your approach is valid for comparative purposes. In many cases, after looking at the instrument for a minute or so, you might decide, for example, that windspeed varies between 15 and 20 km/hour, or is 20–25 km/hour with gusts of up to 50 km/hour. Before beginning to make a comparative series of wind measurements, measure the wind speed at the same spot every minute for ten minutes and see how

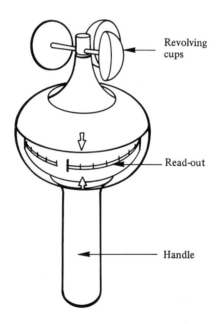

Revolving cups

Read-out

Handle

Figure 3.5 *Hand-held anemometer*

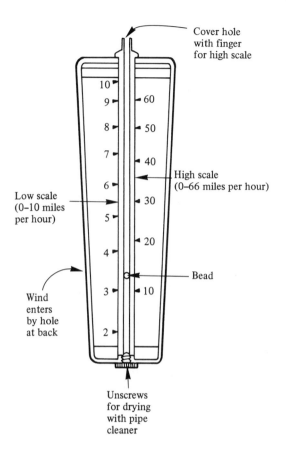

Figure 3.6 *'Dwyer' windmeter (supplied by Griffin)*

variable it is. You may decide a single measurement at each place is adequate, or that an average of several readings would be more meaningful, or that it is the sort of day when wind speed is so erratic that useful measurements cannot be made. You may also decide that it is interesting to measure wind at each sampling point at (say) three different heights, near ground level, as high as you can reach and half way between (select measured heights). If you decide to measure at only one height, it must always be the same one.

3.12 HUMIDITY AS AN ECOLOGICAL FACTOR

Transpiration increases as humidity falls. A major factor which affects humidity is, of course, the weather, but on a field course attention is more likely to focus on microclimatic differences within a small area. Plant species adapted to grow in low humidity tend to be xeromorphic (3.15).

A whirling hygrometer (figure 3.7a) is a type of wet and dry bulb thermometer. The latter (ordinary non-whirling ones are used in glass-houses) consists of two thermometers. One has moist cloth around its bulb, and evaporation of this water depresses the temperature measurement, relative to the dry bulb of the other thermometer. This difference in readings between the wet and the dry bulb is highest when the air is dry (0 per cent humidity), but reduces as the air humidity rises. A special chart (supplied with the instrument) is used to interpret the difference in humidity. This is usually expressed as a percentage relative to the water content of air saturated with water vapour, at the same temperature, and is termed 'relative humidity'. If the air is saturated, as it might be, on occasion, in tropical jungle or a glasshouse, the relative humidity equals 100 per cent.

An alternative means of getting an indication of relative humidity is to use an atmometer (home-made version, figure 3.7b). This measures not humidity *per se*, but the rate of evaporation (affected by wind as well as relative humidity) and this reflects relative humidity sufficiently well to provide a biologically useful alternative. Where relative humidity is high, rate of evaporation is low, and *vice versa*.

Exercise 3.7. Studying air humidity and evaporation rates

What you need
(a) A whirling hygrometer or atmometer (figure 3.7).
(b) Using a hygrometer: a chart and a wash bottle of distilled water.
(c) Using an atmometer: a bowl of water big enough to submerge the apparatus and a supply of spare filter paper discs produced by a hole-punch machine (all the same size).

Method
Using a hygrometer: make sure that the reservoir of the instrument contains distilled water. Hold the hygrometer above your head and whirl it like a football rattle. Read the thermometers, and whirl again. Continue until each thermometer gives a constant reading and find the relative humidity using the chart. Repeat at various places (for example, along a transect).

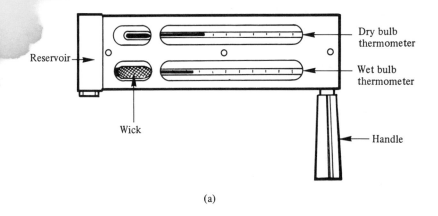

Dry bulb thermometer

Reservoir

Wet bulb thermometer

Wick

Handle

(a)

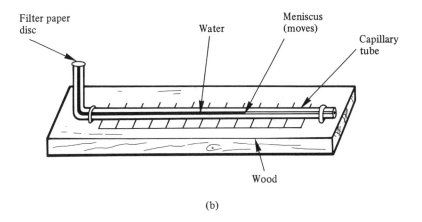

Filter paper disc

Water

Meniscus (moves)

Capillary tube

Wood

(b)

Figure 3.7 *Studying humidity and evaporation rate: (a) whirling hygrometer, (b) home-made atmometer*

Using an atmometer: always record at the same height above ground level, and ensure that the tube is full (just clear of the 'open' end). One way of doing this is to submerge the apparatus in a small bowl of water, replacing the filter paper disc before re-use. Record rate of movement of the meniscus (mm per minute) as an expression of evaporation rate.

It could be interesting to make simultaneous measurements of humidity and evaporation rates at each sampling point.

In your write up

1. How is humidity or evaporation rate related to (i) wind exposure, (ii) altitude, and (iii) air temperature?
2. Why do woodlands have a more stable humidity than open habitats?
3. Usually, woodlands have a higher humidity than surrounding fields. When might it be lower?
4. Small plants in woods are protected from low humidity, and they lack xeromorphic characteristics (3.15). Why are xeromorphic characteristics not only irrelevant but a disadvantage in a woodland?
5. If you have measured both humidity and evaporation rates, see how well they are correlated (8.7, 8.8 and 8.9).

3.13 SLOPE AS AN ECOLOGICAL FACTOR

1. Drainage. Steep slopes are less likely to be waterlogged than level ground.
2. Leaching. Washing out of inorganic ions, many of which are plant nutrients, or, like calcium, affect the pH of the soil (3.17), tends to increase with slope.
3. Instability. On steep slopes, surface instability may make seedling establishment of many species difficult, whilst others such as Parsley Fern (*Cryptogramma crispa*) may be tolerant of this and so they are able to colonise some mountain screes. The establishment of plant cover may be retarded by slope, but successional processes (figure 9.2) may eventually lead to it. In Britain, on scree slopes with a gradient over 35 per cent, colonisation is usually permanently prevented, in which case scree vegetation represents the climax.
4. Erosion. The soil on a slope may be shallow because soil tends to move down the slope as a result of surface drainage (look for erosion gulleys), gravity and frost action. Often, there is deep soil at the foot of a slope.
5. Reduced human interference. Steep slopes are difficult to plough and are more likely to have been left as natural or semi-natural habitats.
6. Aspects. South-facing slopes are warmer than north-facing ones (9.6(b)), and this may affect species composition.

3.14 MEASURING SLOPES BY LEVELLING

Slopes are measured by levelling which consists of determining the difference in altitude between pairs of points a known horizontal difference apart. Map-makers relate altitude to sea level, regarding it as zero. You can give one point (perhaps the lower end of a transect) an altitude of zero, and express other altitudes as values above (+) or below (−) this.

(a) Sections

By making a continuous series of height measurements along a transect, you can draw a 'section' − that is, a diagram of a hillside or sand dune etc. as it would appear cut open. This is done by plotting horizontal distance (horizontal axis) against relative altitude (vertical scale) (figures 3.8 and 3.9). Although you must measure heights and distances in the same units, it is usual to plot them in a section diagram using different scales, exaggerating the vertical measurements, and hence the slope.

Table 3.1 shows a sample sheet for use in drawing a section.

(b) Gradients

$$\text{Gradient (per cent)} = \frac{\text{vertical distance (height difference)}}{\text{horizontal distance}} \times 100$$

For example, if for every metre you go forwards, you go up 0.25 m then

$$\text{gradient} = \frac{h}{d} = \frac{0.25}{1.00} \times 100 = 25 \text{ per cent}$$

Gradient measurements are best made at regular intervals, corresponding to other data. For example, if you have recorded a 1×1 m quadrat every 5 m, calculate the gradient for each quadrat.

Table 3.1   Sample recording sheet for drawing a section of a slope

| Station | Field data | | Data to plot for section | | Gradient (per cent) |
	d (m)	h (m)	Distance (m) (relative to origin)	Altitude (m)	
1 (origin)	0.00	0.00	0.00	0.00	
2	0.75	0.25	0.75	0.25	33
3	0.50	0.25	1.25	0.50	50
4	1.00	0.25	2.25	0.75	25
5	0.90	0.10	3.15	0.85	11

Sample data are plotted in figure 3.8.

49

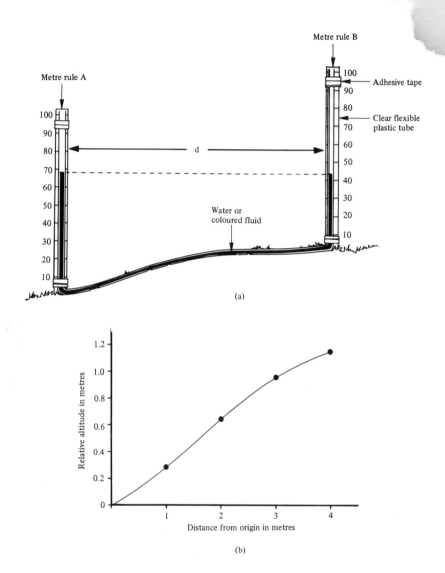

(a)

(b)

Figure 3.8 *Using a manometric level. (a) Home-made manometric level. Example:*
H1 = 68, H2 = 43; h (height difference) = 68 − 43 = 25 cm = 0.25 m. Calculation of
percentage gradient as in figure 3.9. (b) Section of slope using levelling data

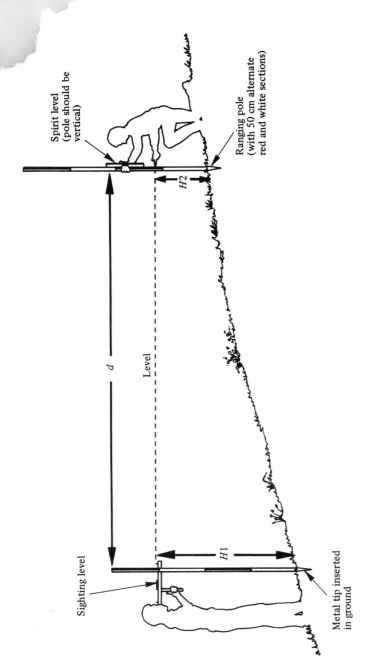

Figure 3.9 *Levelling using ranging poles. To calculate percentage gradient (per cent)* = {(H1 − H2)/d} × 100. *Note that all measurements must be in the same units (for example, all in centimetres or all in metres)*

Exercise 3.8. Levelling using a manometric level

This device (3.8) can be easily and cheaply made. It consists of two metre rules and a length of transparent plastic tubing (about 1 cm or more in diameter) obtainable from aquarium, motor car accessory or home-brew suppliers. To level a series of 1 × 1 m quadrats, the tube needs to be 3 m long, although larger versions to level longer distances can be made by using a longer tube and replacing the metre rules by longer calibrated pieces of wood. Tube length = 2 × length of uprights + maximum distance to be levelled. A plumb-line attached to each upright will give greater accuracy. The tube is partially filled with water (adding ink or some other coloured dye to the water helps) such that when both uprights are placed on level ground, the fluid reaches half way up each scale. To level a section of a transect (such as a 1 × 1 m quadrat), hold each upright vertically at opposite ends of the section, and read the heights of liquid in each side and from this calculate the height difference (h). Measure the distance (d) horizontally to calculate the gradient.

Exercise 3.9. Levelling using ranging poles

What you need
(a) Two ranging poles.
(b) A long measuring tape (say 30 m).
(c) A sighting level.

This method (figure 3.9) is more suitable for levelling longer transects. It is the one used by surveyors. Both ranging poles are stuck in the ground vertically and one recorder stands by the lower pole and looks at the other pole through the sighting level, tilting the level up and down until the bubble visible inside the instrument corresponds to a line. This indicates that he is sighting along a perfectly horizontal line. The height difference (h) can now be calculated (figure 3.9), and you should proceed as in exercise 3.8.

3.15 SOIL MOISTURE AS AN ECOLOGICAL FACTOR

Moisture content of a soil at a given time depends on

(1) Rainfall over the previous weeks and months.
(2) Drainage (function of slope, proximity to streams etc.).

52

(3) The soil's water-holding capacity (related to clay and organic matter content).

(4) Soil depth.

Summer drought is much more common in Britain than is often thought. Cotswold limestone pasture, for example, is often dry in July and August because the soil is well drained (limestone is very porous), the soil is shallow, rainfall is often low, temperatures are high and the hill tops are exposed to (desiccating) wind. Many of the native plant species grow mainly in spring and autumn, and suffer the effects of water shortage (moisture stress) in summer. They may survive because of xeromorphic adaptations, which reduce transpiration and/or allow the storage of water. These include thick fleshy (water-storing) leaves, hairiness (reduces transpiration), reduced stomatal density (reduces transpiration) and several others. On the other hand, marsh or aquatic plants rarely suffer moisture stress, and lack these adaptations. The degree of moisture varies from place to place over short distances, and thus the extent to which species of plants vary in their adaptations to cope with it also vary.

Exercise 3.10. Determining the moisture content of soil samples

The most obvious factor affecting soil moisture content is rainfall. The values that you shall obtain in this exercise are for particular soils on a particular day at a particular time. If, however, you collect samples at various places at the same time, you can compare them, and this may be very useful. There is no point in doing this exercise during or just after heavy rain, when most soils may be fairly saturated, nor after a prolonged drought, when most soils will be dry. Best results will be obtained when it has not rained appreciably for 24 hours but when, nevertheless, most of the soil in the habitat feels moist to the touch.

What you need
(a) Plastic bags (one for each sample).
(b) A waterproof marker pen.
(c) A trowel.

And back at the Study Centre

(d) A top pan balance (preferably to measure down to 0.1 g, but see note (1) below).
(e) A laboratory oven (see note (2) below).

53

(f) Evaporating dishes, (empty!) tobacco tins, and small metal foil trays (as used for small cakes or pies), or similar items (one for each sample).

Notes

(1) If you cannot take a laboratory balance on your field course, a portable battery-operated one, such 'as a 'Soehnle' digital balance, will suffice, even though it may only weigh to 1 g.

(2) A small domestic oven, such as a 'Baby Belling', is quite adequate and is cheaper and easier to take on a field course.

Method

Remove surface vegetation as shallowly as possible, and collect a trowel full of soil, placing it in a plastic bag. Tie up the opening of the bag and label the outside using a waterproof pen with a suitable sample reference number. Try to collect the sample from the organic (or 'A') horizon (3.16).

Back at the Study Centre, weigh a dish or tray for each sample, label it clearly, and enter the weight on a record sheet like table 3.2. Place enough soil to cover the bottom of the tray to a depth of 1 to 2 cm (25 to 50 g) and weigh. Record weight of 'tray + wet soil' on the record sheet, and calculate weight of wet soil by subtraction. Place the trays in the oven (in order, so that, if the heat makes the labels difficult to read, you can still identify them). Heat at $100°C$ (if you are using a domestic oven, check its calibration and remember $100°C = 212°$ fahrenheit) to constant weight (that is, reweigh every few hours until you get two weights the same). Allow sample to cool before weighing to avoid damaging the balance. Usually, heating for 24 hours, or even overnight, achieves this. Record the final weight ('tray + dry soil') on the sheet. Calculate the weight of dry soil

Table 3.2 Sample record sheet for estimation of soil moisture content

Sample no.	Weight of (g)						Percentage water content
	a Tray alone	b Tray + WET soil	c Tray + DRY soil	d WET soil (b − a)	e DRY soil (c − a)	f Soil water (d − e)	$\frac{f}{d} \times 100$
1	39	90	82	51	43	8	16
2	90	126	116	36	26	10	28
3	74	120	112	46	38	8	17

by deducting the weight of the tray. Finally, calculate the moisture content, expressing it as a percentage of the weight of the wet soil.

In your write up
1. Consider the variation in soil moisture content in relation to the distribution of plants with xeromorphic characteristics, and decide if moisture stress could be an important factor in parts of your habitat.
2. Use reference books to find how soil texture may affect a soil's moisture-retaining capacity.
3. The rainfall over a small area will probably be constant and yet soil moisture may vary. Consider your study habitat from this point of view, and suggest explanations for any variation you have found.
4. Find the meaning of the words 'xerophyte', 'mesophyte', 'hydrophyte' and 'halophyte'.

3.16 SOIL ORGANIC MATTER AS AN ECOLOGICAL FACTOR

This term includes all soil organisms and plant roots, but the main component is usually humus. The initial stages of decay quickly cause dead organisms, including leaves in particular, and animal faeces to become the brown or black organic substance which continues to decay slowly. Humus is important in a soil as a reservoir of nutrients, which are slowly released by decay. It also provides cation exchange capacity (9.2) and water-retaining capability.

A soil has a layered structure or soil profile which can be examined by digging a hole (soil pit). The profile can usually be divided broadly into the A, B and C horizons. The A or organic horizon contains humus and is usually darker in colour than the lower layers. It may be subdivided; for example, in some cases a surface litter or A0 horizon containing mainly still recognisable leaves etc. can be distinguished from an underlying A1 horizon. Most (but not all) biological activity is confined to the A horizon, and, unless you are making a special study of profiles, soil samples should be collected from it, avoiding the inorganic B horizon beneath. The C horizon consists mainly of large bedrock fragments. You must consult other books (for example, Geography texts) to find out more about this big subject.

If a soil has a high organic matter content, it may imply accumulation resulting from low microbial activity, as in a peat bog (1.4(b)). Numerous soil animals, such as woodlice and earthworms, feed on dead organic matter (chapter 6), and are very abundant in leaf litter and other organic soils (but not peat bogs), in which they mix humus into the lower layers.

Exercise 3.11. Determination of organic matter content

What you need
(a) Plastic bags and marker pen.
(b) Metal trays or tobacco tins (not the lids – they have rubber seals which burn).
(c) A balance (see note 1, exercise 3.10).
(d) A muffle furnace, bunsen burner, camping stove or domestic oven.

Method
Collect soil samples from the A horizon after removing surface litter, excluding as many roots as possible, and put in labelled plastic bags. You may use the same samples as used to estimate water content (exercise 3.10). Alternatively, you could dig a soil pit, and sample from measured depths. In any event, the samples must have been dried to constant weight at a temperature not exceeding $100°C$ (in a fashion resembling that of exercise 3.10) before proceeding, using a recording sheet which is a modified version of table 3.2. It is convenient, in fact, to determine moisture content and then organic matter in the same samples. Weigh the tray and then tray + *dry* soil (*b*). Heat the tray very strongly using a bunsen burner or camping stove, and the organic matter burns, producing smoke (a possible problem in a laboratory). Continue heating to constant weight. Determine the constant weight accurately. To avoid damaging the balance (especially a plastic portable one), the trays must be allowed to cool (preferably in a desiccator) before weighing. This is even more important than it was in exercise 3.10, since higher temperatures are involved. The final weight (*c*) is the soil without organic matter, and loss in weight 'on ignition' (*b*−*c*) gives the organic matter content, which should be expressed as a percentage of the original (dry) weight of the sample. A domestic oven turned to maximum (usually about $250°C$) can be used, but this takes longer (overnight at least). Many Field Study Centres have a muffle furnace which heats the samples at a temperature of $450°C$. This is the best apparatus for the task. Heating soil overnight in a crucible at this temperature is usually sufficient.

3.17 SOIL pH AS AN ECOLOGICAL FACTOR

Most crop plants flourish best in a soil pH of 6.5 (the pH of the cytoplasm of cells). This is also the optimum pH for many of the microbes which contribute towards soil fertility (by recycling nutrients), and earthworms (also important in maintaining a fertile soil) are absent from acid soils.

Even limestone and chalk soils rarely have a pH of over 8 in Britain. These soils have plenty of microbial activity and soil animals, and are usually fertile. The high pH means that iron tends to be in its insoluble Fe(III) form, and so many plant species suffer from 'lime-induced chlorosis'. Their leaves are yellowish because of iron deficiency. This is mainly noted in introduced agricultural species (for example, grasses such as Cocksfoot, *Dactylis glomerata*), since native species, such as Hoary Plantain (*Plantago intermedia*) and the Upright Brome (*Bromus erectus*) are adapted to these conditions of high pH. Such plants are called 'calcicoles', and, amongst other things, may be adapted to cope with low iron availability. Acid soils, as well as a soil microbe problem, have iron in the soluble Fe(II) form, so much so that many species (especially calcicoles) in such acid soils suffer from iron toxicity, and similar problems, like aluminium toxicity. Plants adapted to survive in acid soils cannot survive in high pH ones, and are known as 'calcifuges'. Examples include the plants of heather moors, such as the Common Heather (*Calluna vulgaris*) and peat bogs, like the Bog Moss (*Sphagnum* spp.) and the Bog Aspodel (*Narthecium ossifragum*). Calcifuginous garden shrubs, such as *Azalea* spp. and *Camellia* spp., do not thrive in calcareous soil unless 'sequestered' iron is applied to their roots by a gardener.

Soil pH may be reduced in British soils because the high rainfall tends to leach out ions (including calcium – often vaguely referred to as 'lime'). This may be evident on freely drained slopes.

Soil pH measurements are often difficult to interpret. If all your values are within one pH unit, you should be content with concluding that soil pH does not vary significantly. The main thing they may tell you concerns the origin of the soil. If the value is over 7, this suggests that the parent rock is limestone (or serpentine) in origin. Millstone Grit soils of the Pennines, and Granite soils of some mountains tend to have a pH below 5. The Bog Moss (*Sphagnum* spp.) exerts its role as the dominant species in a peat bog partly by creating acidity through ion exchange. The most useful situation where you could measure pH is on either side of or across a geological boundary (9.6(d)). The pH will change as you cross the boundary, and you can attempt to correlate this with vegetation. Which species seem to be calcicoles and which are calcifuges? Sometimes there are calcicoles on a roadside even when it crosses acid moorland because limestone has been used to make its foundations.

Exercise 3.12. Measurement of soil pH using indicator

What you need
(a) A trowel.
(b) Either pH-testing kit in a carrying case

or the following

(c) Test tubes, as long as possible, and rubber bungs to fit, or special soil-testing glass tubes.
(d) A small bottle of pH indicator.
(e) A BDH soil indicator chart.
(f) A small bottle of barium sulphate.
(g) A spatula.
(h) A test tube brush.
(i) A container, such as a small bucket, to carry your apparatus.

Commercially available kits have special tubes with a bung each end (easy to clean) and calibrations etched on the surface to help you measure the proportions of reagents. These can be obtained separately but pH can also be measured using ordinary test tubes; make marks with a waterproof pen to give similar proportions. Take extra supplies of distilled water, barium sulphate and indicator with you into the field.

Method (figure 3.10)
This assumes you use the type of tube supplied in the kits, but you can use test tubes in a similar fashion. Place 2 to 3 cm depth of soil in the base of the tube, and then about 1 cm depth of barium sulphate (this will flocculate the clay, and so help it to settle). Make up to the first mark with distilled water (10 cm from base of tube). Add 2 cm depth of indicator (to the second mark). Place the bung in the top, and invert the tube repeatedly until the soil and the solutions are well mixed. Prop up the tube, and leave the contents to settle. Compare the colour of the solution above the settled soil with the chart (consider the *shade* – is it yellow-green, green or blue-green, for example – rather than its intensity).
Wash out tubes carefully (use a brush and some distilled water), before re-using tubes.

Exercise 3.13. Measuring soil pH using a pH meter

What you need
(a) A portable pH meter with a robust field probe (Philip Harris supply a particularly suitable one which is sealed and requires little maintenance). Take a spare battery.
(b) Small beakers or jars.
(c) A pH 7 buffer solution (conveniently made up from special tablets). Make up fresh and take it in a small, wide-mouthed bottle.

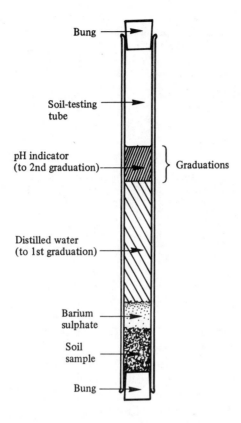

Figure 3.10 *Measuring soil pH using indicator*

(d) A wash-bottle of distilled water, and an additional supply of distilled water to refill it.

(e) A trowel.

Method

Collect a sample of soil, and put about 2 cm depth in a beaker or jar. Make up to a total of 5 cm depth with distilled water, and mix into a paste. Plug the pH probe into the meter, and turn the control to the 'battery test' position. If this is satisfactory, turn the control to the 'on' position, and place the probe in the pH 7 buffer (the standard). When the needle has steadied, adjust the meter (if necessary) to read pH 7. Rinse the end of the probe (squirt water on to it from the wash-bottle), and place it in the soil paste. When the needle is steady, read off the pH.

Rinse the probe again before using with another sample.

You may prefer to use the meter at the Field Study Centre, using soil samples collected from the field.

4   Measuring Environmental Factors in Water

4.1 WATER AS A HABITAT

The previous chapter opened by contrasting the land and the ocean as habitats (3.1). The ocean provided the more stable environment because of the way the properties of a large mass of water differ from those of air. Freshwater habitats share some of this stability, but as much smaller volumes of water are involved, to a lesser extent. Freshwater is also, of course, of much lower salt concentration and hence osmoregulation assumes a greater importance than in marine organisms.

Freshwater habitats provide plenty of variety – some have running water, while in others the water is stagnant. Running water has a higher oxygen content because of several factors that include its lower temperature, the absence of humus which reduces the amount of oxygen used for oxidation of the products of decomposition, and as water flows over stones or waterfalls bubbles of air are trapped and some of this gas dissolves. Animals that live in running waters often have adaptations that allow them to attach themselves to the substratum. There are very few rooted plants in fast-moving waters, and the substratum is mainly rocks and stones because soil is washed away by the current.

Stagnant water differs in several ways, and can be found in, for example, pools, puddles, ponds, canals and ditches. Whereas the temperature of running water remains fairly constant because of the continual mixing, stagnant water undergoes a diurnal and annual variation. The stillness of the water means that dead plant and animal material sinks to the bottom and provides food for animals as well as minerals for plants and phytoplankton. In the spring there may be so much phytoplankton present that the water looks green. As the material decomposes, oxygen is used up and methane gas may accumulate as oxygen runs out, so perhaps not all bubbles that may be visible are bubbles of oxygen from photosynthesis.

Important safety note
Polluted water is particularly interesting to study, but remember it may contain pathogenic microbes and toxic chemicals which could harm you. Wash your hands as soon as possible after working in such places (have a bowl of water with added disinfectant with you in the field), and certainly before eating. Avoid altogether obviously extreme kinds of organic pollution, such as where a whole stream is highly turbid and evil-smelling. Treat any kind of pollution with respect: Wearing rubber gloves is recommended. Be especially careful not to fall into polluted water.

4.2 DISSOLVED OXYGEN AS AN ECOLOGICAL FACTOR

The level of oxygen is high in fast-moving waters, simply because of the shear turbulence (in which water gets mixed with air), but slower-moving waters may have high levels because of an abundance of plant life. Aeration is affected by the depth of the water as well as the character of the stream bed and the frequency of waterfalls etc. Shallow water that runs over a stony stream bed will have more oxygen than water that runs at the same speed but is half a metre deep with a muddy bed.

Oxygen levels often indicate the 'quality' of freshwater. A common form of pollution is sewage, which increases the rate of fungal and bacterial growth, which in turn leads to deoxygenation of the water and the death of animal life (4.12). Well oxygenated water, by contrast, has a high species diversity (7.5 and 7.6).

If the percentage saturation is 90–100 per cent then it can be safely assumed the water is not polluted, between 50 and 90 per cent the water is of fair quality, while saturation levels below 50 per cent indicate the presence of pollution. However, when taking oxygen measurements the time of day must be borne in mind because saturation can fall overnight when photosynthesis stops, or when the temperature rises as the higher the temperature the less oxygen water can hold in solution. However, the very highest oxygen levels are often recorded in polluted streams that have lots of Cladophora (blanket weed), and the water may be supersaturated to as much as 165 per cent.

Aquatic animals have to be able to cope with lower oxygen levels than that in air, and so generally they tend to be slower moving than land animals because there are only 6 cm^3 of oxygen in 1 dm^3 of water compared to 200 cm^3 in 1 dm^3 of air. If water has toxic pollutants then it may also have a low oxygen level, so in order to obtain enough oxygen for their needs animals such as fish may have to pass more water than usual over their gills. As the water passes over the gills the poisonous pollutants are absorbed, so the lower the oxygen level the larger the dose of pollution suffered by the animal.

The effects of organic pollution can be measured by working out the Biological Oxygen Demand (BOD). The higher the BOD then the more polluted is the water. Where BOD is high and oxygen levels low, then there will be little time for recovery before more of the pollutant is added to the water. This water may then contain only those animals which have a low oxygen demand (for example, Tubifex). The absence of fish which are extremely sensitive to low oxygen levels is a good indicator that the water quality is poor, but remember that fish are difficult to catch.

Oxygen levels can be measured by two methods, using an oxygen electrode, or by the Winkler titration (described in chemistry textbooks).

Exercise 4.1. Measuring oxygen levels using an oxygen electrode

What you need
(a) A voltmeter.
(b) An oxygen electrode, spare membrane and gel.
(c) An amplifier and batteries (including spare ones).
(d) A temperature correction probe.
(e) A 250 cm^3 beaker and wash-bottle of distilled water.
(f) Sodium dithionite (optional).

Method
Before use, a delicate membrane must be fitted to the probe, enclosing some special gel. Once prepared, the membrane must be kept wet. Figure 4.1 shows a convenient way of transporting it in the field. The probe is plugged into a battery-operated amplifier which is attached to a voltmeter. (Some more expensive apparatus may have the voltmeter and amplifier enclosed in the same case.) Some models also need a temperature correction probe.

Before use the probe may be calibrated (but see below) as follows.

(1) Prepare water which is saturated with oxygen by bubbling air through it for several hours. Treat this as containing 100 per cent dissolved oxygen.
(2) Prepare water with zero dissolved oxygen by either of the following methods.
　(i) Boiling and cooling it (*note*: it will soon absorb oxygen from the air and must be freshly prepared).

63

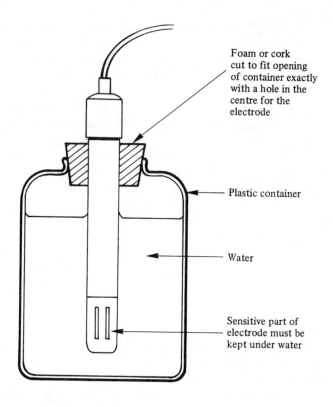

Foam or cork
cut to fit opening
of container exactly
with a hole in the
centre for the
electrode

Plastic container

Water

Sensitive part of
electrode must be
kept under water

Figure 4.1 *Transporting an oxygen electrode safely*

(ii) Adding a few grams of sodium dithionite which rapidly absorbs oxygen (*note:* remember to wash the dithionite off the probe before measurements are taken).

Place the probe in the 100 per cent saturated water, and adjust the zero control to give the biggest deflection possible. Use the two solutions to prepare a calibration graph, assuming that the readings are linear, in a similar fashion to that described for the temperature probe (exercise 3.4).

This is not a very convenient method in the field, where the simplest method is to regard the atmosphere as of constant oxygen concentration. Hold the probe in the air and adjust the zero control to a full-scale deflection (or a suitable readout on the voltmeter which must be the same prior to every reading). The readings will be in volts. This could be converted to percentage saturation as described above, but to make comparisons between oxygen levels of different water samples this is not essential.

The procedure for using the probe in the field is similar to adjusting a pH meter with a buffer (exercise 3.13). Assemble the apparatus, take a water sample and place it in a beaker. Before taking an oxygen reading, hold the electrode in the air and adjust the zero control to give a full-scale deflection. Next, place the probe into the water sample and then, once the reading is stable, which may take a few minutes, record the voltmeter output. This gives an indication of the level of oxygen in the water as compared to air (figure 4.2).

Take a few different samples from the water and take more readings to make sure that the results are consistent. Alternatively, measurements can be taken over a day to investigate any changes due to photosynthetic activity using a datamemory (2.10). Consider how you could use this apparatus to compare the BOD of freshly collected water samples from different sources.

4.3 pH AS AN ECOLOGICAL FACTOR IN AQUATIC CONDITIONS

The pH of water depends on a variety of factors, an important one being the nature of the rock it flows over. A high pH of water in a stream or pond may indicate that it has drained from a limestone area, and be associated with a large mollusc population. These animals, which require calcium to form their shells, are often absent from acid water. Limestone valleys are typically 'dry', and the presence of a stream or lake (such as Malham Tarn) indicates that the actual stream or lake bed is not calcareous but composed of some other, more impervious, rock. It is the source of the water which matters. The level of carbon dioxide can also affect pH, because at night there is a build-up of this gas as no photosynthesis takes place, so pH falls slightly.

The pH of water is important because many biological activities can occur only within a narrow range, so any variation from the range could be fatal to a particular organism.

The easiest way to test pH is to add some universal indicator to a 10 ml water sample, or a pH meter can be used (exercise 3.13).

4.4 TEMPERATURE AS AN AQUATIC ECOLOGICAL FACTOR

As already mentioned, the solubility of oxygen decreases with increasing temperatures, and this leads to environmental stress. Water at high altitude will, however, contain a range of organisms that are adapted to lower temperatures.

65

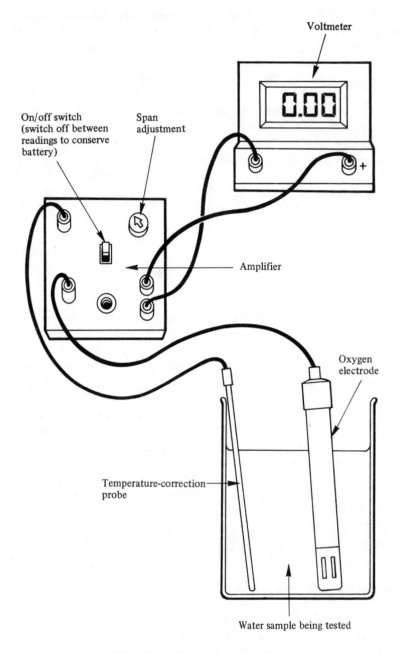

Figure 4.2 *Taking an oxygen reading*

There are, of course, seasonal and diurnal changes, and if a maximum and minimum temperature can be recorded then some idea of the range of temperatures between which the organisms have to live can be obtained. In deeper waters there may also be some thermal stratification, the uppermost regions being warmer with the animals collecting in the cooler, more oxygenated deeper regions. There may also be some horizontal variations as some areas may be shaded by overhanging trees.

Generally, over the course of a day the uppermost water layers may experience a slight temperature increase, while the deeper regions may remain at a constant temperature throughout.

Exercise 4.2. Measuring the temperature in water

What you need
(a) A comparator.
(b) A temperature probe.
(c) Some extension poles for probe.
(d) A waterproof pen.
(e) A datamemory (optional for 24 hour recording).

Method
Field temperature measurements can be taken by placing the probe directly in the water and taking the reading of the scale. When making a series of determinations, all recordings must be made at constant depth.

To do this, measure the distance (for example, 0.5 m) from the sensor at the end of the probe back along the pole, and then mark this with a waterproof pen. Now all that needs to be done is to lower the probe into the water until the mark touches the surface of the water.

If you are sampling in a small pond then it may be worth while recording temperatures at different levels, using a similar method (with or without a datamemory, 2.10) to that of exercise 3.5. To make sure that the readings are made at precisely the same depth and in the same place, a 'rig' can be assembled using clamp stands and this will hold all the probes in the same position for as long as necessary.

Calibrate your meter as in exercise 3.4.

In your write up
1. Explain why the temperature changes during the course of a few hours, or over a day.
2. Discuss what effect this might have on living organisms.

3. What main factor determines the water temperature? Does this factor have the same effect along a stream, or all over a pond?

4.5 MEASURING TURBIDITY, TOTAL AND SUSPENDED SOLID CONTENT

Turbidity is the name given to the clarity of the water, which is affected by the amount of suspended solids present in it. High turbidity often accompanies organic pollution.

The more turbid the water is, then the greater the level of organic pollution such as sewage. If the water contains a large amount of suspended solids, then this is detrimental to the animals that have gills, as the particles gather on the gill filaments, block them up and reduce the surface area for gas exchange. As this continues, the animals will die as they are unable to obtain sufficient oxygen to supply their metabolism.

Turbidity reduces the depth to which light can penetrate, and hence reduces the growth of plants. Since these are the primary producers of the ecosystem (1.4) and the main source of oxygen, this also has a detrimental effect on the animal life. Sometimes the main source of turbidity is the very plant life itself, as with algal blooms (4.12).

Exercise 4.3. Measuring the turbidity of water

What you need
(a) A glass tube, 2.5 cm diameter, 1 m long.
(b) A piece of plane glass glued to the bottom which has a black cross painted on with lines 1 mm in width.
(c) A 250 cm^3 beaker.

Method
A water sample must be obtained and this is then poured into the tube until the cross can no longer be seen when looking straight down the tube. Next, measure the depth of water in the tube and record the measurement. This can be repeated with water from different ponds, or different regions along a stream which receives effluent at one point — for example, a farm drainage outfall pipe.

Results
The measurements can be classified as follows

Good waters	above 600 mm
Satisfactory	about 300 mm
Poor waters	less than 100 mm

68

Another method is to use a Secchi disc which is divided into four alternate black and white sections. It is lowered on a piece of string into the stream or pond until the sections can no longer be seen. The point where the string enters the water is marked in some way (by holding it or having coloured thread tied to the string at measured intervals) and the disc is removed from the water. The distance between fingers (or the lowest coloured thread left exposed) and disc is then measured and this is the depth to which the disc was lowered. The disc is most easily used from a boat or a bridge.

Exercise 4.4. Measuring the suspended solid content

What you need
(a) A filter funnel.
(b) Filter paper (of known weight).
(c) A measuring cylinder.
(d) The water sample.
(e) A 250 cm^3 beaker.
(f) A balance.

Method
Take a sample of water of known volume and pour it through a filter funnel, letting the water drain into a beaker. Remove the filter paper and let it dry naturally. Weigh the filter paper again and work out the weight of the sample.

Results

$$\frac{\text{Weight of solid (g)}}{\text{Volume of water} \times 1000} = \text{grams/litre}$$

Exercise 4.5. Measuring the total solid content

What you need
(a) The water sample.
(b) A conical flask and a rubber hung.
(c) An evaporating dish (of known weight).
(d) A water bath.
(e) An oven.

Method

Place the water sample in the conical flask and shake very thoroughly. Pour the sample into the evaporating dish and weigh it, and work out the weight of the water sample. Place the dish over a water bath and evaporate to dryness. Then place it in an oven at 105°C. When thoroughly dry, weigh the dish and work out the weight of the solid.

Results

Weight of dish   =
Weight of dish and sample   =
Weight of sample   =

Weight of dish and sample after drying   =
Weight of sample   =

$$\text{Percentage solid content} = \frac{\text{Weight of sample before drying}}{\text{Weight of sample after drying}} \times 100$$

In your write up

1. What is the solid matter composed of?
2. Using these results, describe the effect of the presence of a large amount of solid material in the water.
3. Is the animal population of the waters that you have studied showing any signs of suffering from too much solid material?

4.6 CONDUCTIVITY AS AN ECOLOGICAL FACTOR

The level of conductivity in water gives a good indication of the amount of ionisable substances dissolved in it, such as phosphates, nitrates and nitrites which are washed into streams and ponds after fertiliser is applied to surrounding fields or are present in effluent from sewage-treatment installations. The sodium chloride and other salts in seawater give it a higher conductivity than freshwater. In a rock pool, as water evaporates solute concentration – and with it conductivity – rises. Most organic substances present in sewage etc. are not ionisable and so do not affect conductivity (many are present as solid, suspended particles). Once decay begins, however, they start to break down to inorganic, ionisable sub stances (nitrate, nitrite, phosphate etc.). In unpolluted waters the conductivity increases by approximately 3 per cent with every 1°C rise in temperature, so temperature must be noted, and corrections made for comparative purposes (relative to the sample with the lowest temperature).

The easiest way to measure conductivity is by using a conductivity meter and probe. The probe can be placed directly in the water or a sample can be taken, placed in a beaker and the probe placed in that. The reading can then be taken from the display on the meter.

4.7 PHOSPHATE, NITRATE AND NITRITE AS ECOLOGICAL FACTORS IN FRESHWATER

Phosphate usually gets into water from detergents, but some phosphate is excreted in urine. Detergents with a complex phosphate base are often added to sewage and this then flows into the water. The phosphate then acts as a fertiliser for the growth of plants and animals.

Nitrate usually gets into the water from field drainage, usually directly into the nearest stream. This run-off brings with it a high content of nitrogen from fertilisers.

A concentration of nitrate greater than 20 mg per litre is a health hazard because it is reduced by bacteria in the gut to nitrites which oxidise haemoglobin to methaemoglobin, and so destroy the ability to carry oxygen. In natural conditions a level of nitrate of 1 ppm is usual, and it rarely exceeds 10 ppm.

Nitrite is oxidised to nitrate by bacteria in the water as part of the nitrogen cycle, and it is easier to accurately measure this than nitrate.

The easiest way to measure these chemicals is by using a Nessleriser, which is similar to a Lovibond comparator, together with a Nessleriser disc. (See instructions for use supplied with the equipment.) The method is colourimetric and can be carried out in the field or with samples returned to the laboratory. The necessary reagents are supplied in tablet form and are easy to make up accurately. Remember to take a pestle and mortar with you into the field to crush the tablets, which makes them easy to dissolve. You must take a supply of distilled water with you too.

There are also dip sticks available that work in a similar way to clinistix; these are simply placed in a water sample and then removed, and after 30 seconds the colour change can be compared with a chart on the side of the tube, which indicates the concentration of say nitrate in the water sample.

4.8 SALINITY

Chloride ions are often present in badly polluted waters and in tidal reaches of streams, where conductivity readings can be misleading. The usual method of estimating their concentration is by volumetric analysis using silver nitrate which, of course, has to be carried out in the laboratory. This method is described in standard chemistry books. Any increase in chloride ions in water that is usually classed as 'fresh' will have a drastic effect on the plants and animals as it will affect their osmoregulation abilities. Freshwater plants and animals normally have a lower (more negative) osmotic potential than the surrounding water. If the salt concentration of the surrounding water rises, then its osmotic potential falls (more negative). If it falls below that of the organisms, then water moves out of their cells into the surrounding water, which of course leads to the death of the organisms.

Exercise 4.6. Measuring the salinity using Quantab 1177 chloride titration papers (available from Philip Harris)

What you need
(a) A beaker.
(b) The water sample.
(c) Quantab 1177 Chloride Titration Papers.
(d) The calibration table (supplied with the papers).

Method
Place the water sample in the beaker, then place the capillary tube that contains silver dichromate in the sample. The water rises up the tube because of capillarity, and a band of white silver chloride forms if chloride ions are present. There is a yellow band at the top which turns blue, and a reading must be taken as soon as the band changes colour; this takes about 10 minutes. The length of the white band must be read about 2-3 minutes after the blue colour appears. The size of the band can be read off the calibrations that run along the side of the tube. The concentration of chloride ions can then be worked out using the table that is provided with the tubes.

4.9 CALCIUM AS AN ECOLOGICAL FACTOR IN FRESHWATER

It is important to determine hardness. The level of hardness does not vary much in a river unless there is a significant level of pollution. A good indication that water is hard is an abundance of gastropods, gammarids and bivalves which are absent from soft water. Calcium can have a dramatic effect on a freshwater community, the dividing line being 20 mg/litre. As hardness decreases, pH increases (4.3) and mayfly and stonefly appear. Tablets for estimating hardness are available from suppliers and are very simple to use.

4.10 MEASURING DEPTH

The depth of a river, stream or pond can be measured. However, care must be taken as it is easy to lean out too far and fall in. Ideally, measurements should be taken from a bridge or a boat, by using a long pole with measurements marked along it.

4.11 MEASURING FLOW RATE

Expensive flowmeters can be used to measure the flow rate of a stream. However, a simple method of doing the same thing is to use a 'Pooh' stick, a stopwatch, and a measuring tape. The tape is laid down along the bank, the 'Pooh' stick is placed in the water at 0 metres, or equivalent, and using the stopwatch, the length of time it takes for the stick to travel one metre or so is measured. Repeat several times and take the mean.

Another method is to use a Rubberbag meter (after Hynes, *The Ecology of Running Waters*, Liverpool University Press, 1970), which is cheap and fairly accurate and is more useful for microhabitat work. It consists of a glass tube of known diameter with a plastic bag attached to one end. The meter is put into position with your thumb over the hole. The thumb is removed and water is allowed into the bag for a fixed period of time; the thumb is then replaced, and the meter is removed from the water. Next the water is poured from the meter into a measuring cylinder and the volume noted.

The results can be calculated using the formula below.

$$\frac{\text{Water in bag}}{\text{Cross-sectional diameter of tube}} = \text{Flow rate per unit time period}$$

4.12 POLLUTION AND ITS EFFECT ON FRESHWATER

Probably the most common form of pollutant is sewage. Most is treated before discharge and most organic matter is removed by sedimentation and the action of bacteria in the filter beds. There are two types of sewage systems. One collects waste from land drainage, carrying water straight to the rivers; the other collects waste from households and industry and takes it to sewage works. Sometimes, particularly in the sewage works of large cities, because of the volume of input into the works it is often released before treatment is complete. This may lead to an increase in algal growth because of the large quantities of nutrients, particularly phosphates, being released into the water, so causing eutrophication.

Eutrophication, where water has become heavily loaded with nutrients, can be a natural process and happens particularly in woodland streams which receive fallen leaves; as the leaves rot, minerals are released into the water. Most natural rivers are usually low in nutrients and are known as 'oligotrophic'. The word 'eutrophic' simply means 'nutrient rich'. Many ponds might be described as eutrophic for quite natural reasons, and can be fairly stable ecosystems (1.4). Provided that nutrient input is constant from year to year, detritus food chains will maintain an equilibrium. 'Eutrophication', the process of becoming more eutrophic, can be an ecological villain when it is rapid, usually as a result of human activity including sewage disposal and agriculture. A rapid influx of extra nutrients upsets the balance and throws the ecosystem into chaos. The algae grow rapidly causing an algal bloom, which cuts out the light supply to the deeper regions of water, so the plants below the surface die and decay. This in turn leads to a further increase in nutrient level and a further increase in growth rate until sheer space becomes a limiting factor. The algal bloom represents a massive accumulation of organic matter and as competition for light kills the lower layers of the algal bloom, saprophytic fungi and bacteria are stimulated to grow. These absorb oxygen and so deoxygenate the water, killing fish and all other aerobic forms of life.

Not all algal blooms, however, indicate a sudden influx in external nutrient supply. They can occur when nitrite levels are as low as 0.3 ppm and phosphate levels are merely 0.01 ppm, when, for example, insecticides kill the herbivorous Crustaceans which normally keep the algae in check.

Synthetic detergents pass through sewage treatment works and pollute the water. Detergents release a large amount of phosphate, but domestic sewage is seldom poisonous; it just encourages the growth of different organisms. Levels as low as 0.1 ppm of detergent cause foaming. Industrial sewage often does contain poisons and metals such as copper, zinc, lead and silver accumulate in the tissues of animals such as fish, which are then eaten by other predators such as man. Pollution by animal wastes that drain from nearby fields acts just like untreated urban sewage. Silage also produces effluent which has the immediate effect of polluting the water into which it escapes.

Animals are by far the best indicators of water quality, and this is dealt with in chapter 7.

4.13 DETERMINATION OF BACTERIAL POPULATIONS IN WATER SAMPLES

This is easily done using 'dip slides' marketed by Griffin. The slides are covered by a layer of agar medium and supplied ready sterilised in glass specimen tubes. The dip slide is simply dipped in the water and rapidly replaced in the tube, which is then placed in an incubator until colonies can be counted. The number of colonies gives an estimate of the original size of the bacterial population.

5 Quadrats

5.1 WHAT IS A QUADRAT?

A quadrat is a small area of ground marked out for the purpose of making a detailed description and recording numerical data, and usually acting as a sample of a larger area. As the name implies, a quadrat is usually square or rectangular, but it can be round or take some different form (for example, a point quadrat – see below). They are used to sample objects which do not move, at least during the sampling, such as plants and the more-or-less sedentary animals of the seashore. They could also be used to sample, for example, the size of pebbles on a beach or the distribution of ant-hills in a pasture. Quadrats can be any size, but commonly they are 1 x 1 m or smaller. In a jungle or forest, much larger ones might be used, but they would probably be called 'sample plots'.

Quadrats can be marked out by knocking marker pegs in the ground, but in your kind of work, you will use a quadrat frame (which is often just called a 'quadrat'). These can be bought ready-made but you can make your own more cheaply. Small ones (for example 0.1 x 0.1 or 0.25 x 0.25 m), as used in exercise 2.1, can be made from stiff wire or by soldering four metal rods together, as shown in figure 5.1, which also shows how to make a larger one out of wood. This has holes through which string can be threaded to divide it into 25 squares (5.3(b)). For general use, a size of 0.5 x 0.5 m is particularly suitable. 1 x 1 m quadrat frames tend to be inconveniently large. Folding ones are available, but the simplest solution is to lay 4 metre rules (or other metre long pieces of wood) to form the frame on the spot. Point quadrat frames will be dealt with later.

5.2 RECORDING QUADRATS

(a) Species frequency
Definition: the probability of an individual of a particular species being present in a randomly placed quadrat.

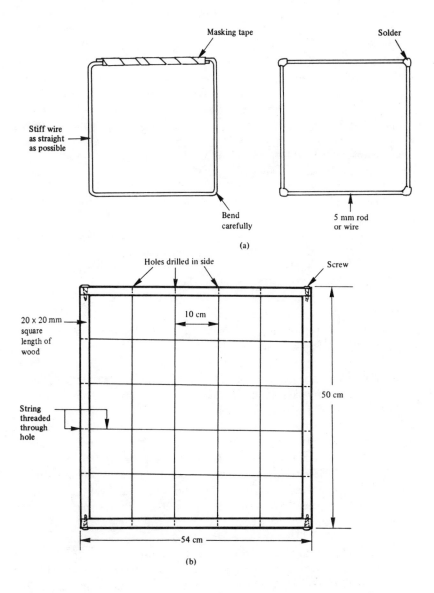

Masking tape

Solder

Stiff wire
as straight
as possible

Bend
carefully

5 mm rod
or wire

(a)

Holes drilled in side

Screw

20 x 20 mm
square
length of
wood

10 cm

50 cm

String
threaded
through
hole

54 cm

(b)

Figure 5.1 *Home-made quadrat frames: (a) wire quadrats, (b) wooden quadrat for
cover estimates*

When recording the frequency of the species present in a study area, it is only necessary to record whether each species is present or absent. You can, of course, treat other forms of quadrat data from a frequency point of view; for example, if you have recorded density, zero means 'absent' and anything else means 'present'. Frequency estimations are affected by quadrat size. Obviously, if your quadrat were so big that it covered the whole study area, every species would be shown to have a frequency of 100 per cent. If it were very small (such as 1 × 1 cm), even with numerous quadrats many species would go unrecorded (that is, they would be shown to have a frequency of 0 per cent). In a comparative study, stick to the same quadrat size and state it clearly when you present your results.

Express species frequency as follows

$$\text{Frequency of species } x \text{ (per cent)} = \frac{\text{No. of quadrats with } x \text{ present}}{\text{Total no. of quadrats in sample}} \times 100$$

(b) Species density
Definition: the number of individuals per unit area.

This was the subject of exercise 2.1, where you probably found estimations were also affected by quadrat size.

(c) Species cover
Definition: the proportion of ground surface occupied by, and overhung by, the foliage of a particular species.

Total cover in a quadrat may exceed 100 per cent. In a wood, for example, you must not only record the herb layer, but also look upwards and consider to what extent the quadrat is overhung by shrubs and trees (3.5 and table 8.1). A similar problem may arise in short turf which has a few plants taller than the rest, especially if the latter have large leaves. Grassland and heath vegetation may also be layered, with much more moss growing beneath the foliage of the higher plants than is at first apparent.

As will be clear from the following section, estimation of cover can be more subjective than that of frequency and density, and yet it is often the most useful, particularly when you are concerned with the species composition of the vegetation as a whole rather than with a few individual species as they occur in different places. In a forest, the species with the lowest density will probably be the tree species (much less than 1 per m^2) because of the great size of the individual organisms. In this case, density is biologically very misleading as the trees include the community's dominant species, and contribute the bulk of the ecosystem's primary productivity (1.4). Another reason for recording cover is that it is very difficult to determine density of species like grasses and mosses which consist of numerous small individuals.

78

5.3 ESTIMATING SPECIES COVER

(a) Subjective estimation of percentage cover
This can be done using a quadrat divided into 25 squares (figure 5.1b). For each species, assess the percentage of the quadrat covered by it. Try to imagine all the plants collected together in one corner. If your quadrat is divided into 25, remember that each small square represents 4 per cent of the total cover. Species accounting for less than 1 per cent cover should be given a nominal value of 0.5. This method must not be hurried, and you must examine every part of the quadrat before recording begins. The correct posture involves having your nose less than 50 cm from the plants, and you need to gently handle the plants to find what lies amongst and beneath their foliage. It is a good idea to have several people record the same quadrat without knowing the others' estimations, and then to compare their results. Practise the technique in this way before doing serious ecological work. Experienced ecologists usually get consistent results. The subjectivity in this procedure is justified when (as is often the case) no other method is appropriate. The quadrats should, however, have been placed according to an objective scheme.

Table 5.1 is a sample recording sheet which also shows how to use the same data to calculate frequency. Collate your quadrat data by calculating the mean percentage cover for each species. Use the Mann–Whitney test if you want to test two lots of quadrat data in a comparison of two sites (2.6).

(b) Estimating cover by point quadrat
A point quadrat frame is illustrated in figure 5.2. It is placed in position according to some objective scheme. A knitting needle ('pin') is dropped through each hole in turn. Record every time that it 'hits' a plant, on a recording sheet as in table 5.2, using the 'five-barred gate' method (four vertical strokes and one horizontal = 5 hits). Move the frame to the next position and repeat the process. Note that, as shown in figure 5.2, the pin may score a multiple hit, and you may get more than 10 hits from 10 pins. One way to position the quadrats is to lay tapes parallel and an equal distance apart across the study area, and record 10 pins every 50 cm, placing the frame at right angles to the tape. This is effectively a regular grid (2.3). You could alternatively place the frame at random points in a grid (2.2). Calculate for each species

$$\text{Percentage cover} = \frac{\text{No. of hits}}{\text{Total no. of pins}} \times 100$$

Table 5.1 Sample recording sheet (species, by percentage cover)

Name: Site:

| Species | Percentage cover of quadrat no. | | | | | | | | | | Cover | | Frequency | |
	1	2	3	4	5	6	7	8	9	10	Total	Per cent	Total	Per cent
Daisy	15	5	4	18	20	3	1	0	0	20	86	8.6	8	100
Buttercup	20	30	31	20	5	5	5	4	4	0	124	12.4	9	90
Dandelion	2	5	5	5	2	2	5	5	18	4	53	5.3	10	100
Self-heal	4	5	0	0	20	30	5	5	0	2	71	7.1	7	70
Bent grass	40	70	80	85	90	70	70	40	65	45	655	65.5	10	100

Table 5.2 Recording sheet for point quadrats

Name of recorder: Site:

Species	Raw data				Individual results		Class results	
					Total hits	Per cent	Total hits	Per cent
Daisy	卌 卌 卌 卌				37	18.5	250	12.5
	卌 卌 卌 11							
Creeping buttercup	卌 卌 卌 卌				51	25.5	500	25.0
	卌 卌 卌 卌							
	卌 卌 1							
Dandelion	卌 卌 11				12	6.0	130	6.5
Self-heal	卌 卌 卌 卌				28	14.0	300	15.0
	卌 111							
Bent grass	卌 卌 卌 卌				135	67.5	1700	85.0
	卌 卌 卌 卌							
	卌 卌 卌 卌							
	卌 卌 卌 卌							
	卌 卌 卌 卌							
	卌 卌 卌 卌							
	卌 卌 卌							
	Total no. of pins				200	100	200	100

The point quadrat method can provide a more objective approach to estimating cover with greater precision than the one described previously, but deciding whether the pin has brushed against a blade of grass can be very subjective. The technique tends to over-estimate cover because of the thickness of the pins. They should be as thin as possible (in theory, infinitely thin). Point quadrats are useful in short turf (such as a closely mown lawn) and where plant cover is sparse. There is obviously a problem in a wood (you cannot get knitting needles long enough!) — although when you have recorded the ground flora, you could raise the frame on poles and peer upwards through the holes, recording cover or sky. The method can also be very difficult in long grass. If in doubt, try it and see how practicable it is in the circumstances. The previous, subjective, cover estimation method can be used anywhere.

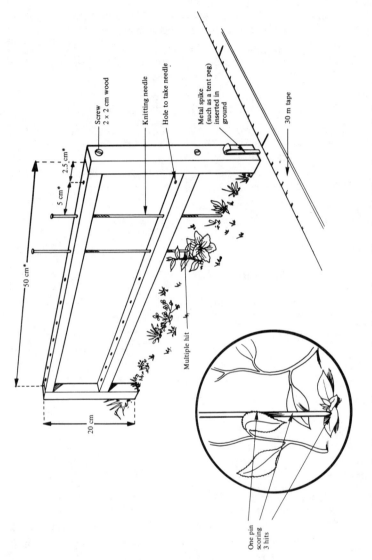

Screw
2 × 2 cm wood

Knitting needle

Hole to take needle

Metal spike
(such as a tent peg)
inserted in
ground

30 m tape

2.5 cm*

5 cm*

50 cm*

20 cm

Multiple hit

One pin
scoring
3 hits

Figure 5.2 *Home-made point quadrat frame (10 pins, 50 cm long). The values marked with an asterisk should be doubled to make a 10 pin, 1 metre long frame*

82

5.4 SUBJECTIVE SCALES: THE DOMIN SCALE

People doing cover estimates in quadrats sometimes agonise over whether a species should be awarded 12.5 or 15.0 per cent. Nobody can be so precise. Subjectivity can be 'tamed' by assigning each species to a broad yet clearly defined category, as illustrated by the Domin cover/abundance scale.

This combines a cover estimate for the species which account for most of the cover, and density (= abundance) for the rest. Award each species a score according to table 5.3. This method enables recorders, with a little experience, to achieve very reproducible results. The main disadvantage is that the scale is non-linear, and so calculation of means for groups of quadrats, and statistical tests (such as Mann-Whitney, scatter diagrams and correlation coefficients) are not possible using the raw data. However, this problem is solved by replacing the Domin numbers by equivalent transformed values (given in table 5.3). If you can cope with the transformation, the Domin scale is probably better than simple cover estimates for much ecological work. The authors' 'ecostat' computer program (2.9) can do this for you.

When using this scale, first consider cover. If it is less than 5 per cent, award a score between + and 4 on density. If cover is greater than 4 per cent, ignore the density column, even if only a single plant is involved, and award a score on cover.

Table 5.3 The Domin cover/abundance scale

Cover (per cent)	Density	Domin symbol	Transformed Domin value
< 5	1 to 2 plants	+	0.04
< 5	3, 4 or 5 plants	1	0.2
< 5	6 to 10 plants	2	0.4
< 5	11 to 30 plants	3	0.9
< 5	31 to 100 plants	4	2.6
5-10		5	3.0
11-25		6	3.9
26-33		7	4.6
34-50		8	5.9
51-75		9	7.4
76-100		10	8.4

Adapted from Bannister, *Journal of Ecology*, Vol. 54, pp. 665-674, reproduced with permission.

Quadrat	1	2	3	4	5	6	7	8	9	10	Mean
Daisy (Domin no.)	5	4	6	4	3	2	7	5	3	4	
Transformed value	3.0	2.6	3.9	2.6	0.9	0.4	4.6	3.0	0.9	2.6	2.45

5.5 POINT QUADRATS PER UNIT AREA

If the vegetation makes point quadrats possible, this approach enables point quadrat data to be combined with the collection of physical, chemical and animal data (chapter 8). You require a 0.5 x 0.5 m quadrat frame and a 0.5 m point quadrat frame. Place the square quadrat frame in position and the point quadrat frame parallel with one side and 10 cm from the end of the two adjacent sides. Record ten pins and move the frame 10 cm along, recording another 10 pins. Repeat until you have dropped 50 pins for the 0.25 m² area, and use these to calculate the percentage cover for each species. Record any species present in the quadrat but not hit either simply as present (frequency), or count the individuals (density).

5.6 MEASURING SPECIES DIVERSITY

'Species richness' simply means the number of species in a habitat and may be applied to plants, animals and micro-organisms. It can be very useful in many situations, but, as many species may be represented by only a few, easily overlooked individuals, species richness tends to get under-estimated.

If a field contains 1000 plants of which 996 are of the same species and four other species are represented by only one individual, you may feel it is somehow misleading to regard this as having the same diversity as another field of 1000 plants in which each of five species is represented by 200 plants. The first field could be regarded as basically having one species and a few freaks, whilst the other is the more diverse. In the second field, should a species whose total population of 200 plants are all in one corner carry the same weight as one whose 200 plants are evenly spread throughout the field? There is, in fact, no simple, precise definition of diversity, but attempts have been made to deal with these problems. One such approach is to use Simpson's Index, an index of diversity calculated from the formula

$$D = \frac{N(N - 1)}{\Sigma n(n - 1)}$$

where D = diversity (Simpson's) Index
   N = total number of individuals
   n = number of individuals of each species
   Σ = sum of.

Sample calculation
The data are from an unused vegetable plot.

Species (no. of individuals)	n	$n - 1$	$n(n - 1)$
Groundsel	45	44	1980
Shepherd's Purse	40	39	1560
Dandelion	10	9	90
Total (N)	95		

$$\Sigma n(n - 1) = 1980 + 1560 + 90 = 3630$$

$$D = \frac{95 \times 94}{3630} = \frac{8930}{3630} = 2.46$$

Problems with plants
Numbers of individuals are readily determined when working with animals, but with, for example, grass or moss, there are numerous small plants packed closely together, and it is often impracticable to count them. In this case, use point quadrat data, regarding hits as individuals. As this is technically 'cheating', be careful to mention that 'the diversity indices were calculated from cover values determined by point quadrat in which hits were regarded as individuals'. You must have plenty of data (team-work is essential). You will under-estimate diversity by this method, but your values for each site will be comparable since they are collected by the same procedure.

Ecologists consider diversity indices must be interpreted with care. In addition to the Simpson's Index described above, there are several others (such as the Shannon–Weiner). One way in which they differ is the emphasis that they give to less frequently encountered species. Simpson's Index is a useful introduction to the subject, but in an advanced study, an ecologist might prefer to describe a habitat using several types of diversity index. Usually, separate diversity indices are calculated for the plant and animal populations in a habitat.

Interpretation

There may be exceptions, but, in general

1. A high diversity index suggests that the community is ancient and ecologically stable, and so likely to remain the same for centuries.
2. A low index suggests that the community is of recent origin or affected by recent changes (usually caused by humans – agriculture or pollution). It is probably unstable and likely to change rapidly.

A recently established lawn has a low diversity, and those who use selective weed-killers aim to keep it so. Some lawns are derived from a meadow when a garden was enclosed, and a high index may imply that it is much older than the associated house. Similarly, the high diversity of Cotswold limestone grassland shows its ancient nature, in contrast to that of recently sown agriculturally 'better' pasture. Spruce plantations are recent creations, and many of the associated species of more natural Norwegian spruce forest may be absent from Britain. Forestry practice – trees planted close together and all of a similar age, without the gaps left by naturally fallen trees – also lowers diversity. The ultimate in low diversity is a monoculture with only one species (for example, wheat). This, the ideal of a modern farmer, is ecologically unstable and requires constant human intervention to maintain it. Cotswold limestone pasture only requires a certain level of grazing to maintain its ecological complexity. Climax Oak wood requires no human attention at all except to be left alone.

5.7 SPECIES ASSOCIATION INDEX

Most apparently homogeneous vegetation is not, in fact, perfectly uniform, but shows pattern, often because the environment is not uniform either. Some parts may be damper than others, and certain species may be positively associated because they tend to grow together in such parts of the site and often get recorded in the same quadrat. Other species may similarly be associated in the drier parts. On the other hand, daisies may rarely ever occur in the same quadrat as the Common Rush, because they normally grow in the drier parts whilst rushes are confined to the damp patches. In this case, the association is negative.

To calculate a value to express such associations, you need frequency data (presence/absence in each quadrat), but you can easily derive this from other types of quadrat data, such as zero cover = absent and any other score = present. The null hypothesis (2.4(b)) — that is, there is no association — is tested by considering the probability of the species occurring in the same quadrat.

For n quadrats and species A and B, let

a = no. of quadrats with A and B
b = no. of quadrats wth only B
c = no. of quadrats with only A
d = no. of quadrats with neither A nor B.

Arrange these in a contingency table, presented here with specimen data (where $n = 100$).

SPECIES A

	+	−	
+	$a = 75$ (68)	$b = 5$ (12)	$a + b = 80$
−	$c = 10$ (17)	$d = 10$ (3)	$c + d = 20$
	$a + c = 85$	$b + d = 15$	$n = 100$

(left label: SPECIES B)

To test for significance calculate chi-squared as an association index. You may have met chi-squared in, for example, genetics, but here it is calculated in a special way.

$$\text{chi-squared} = \frac{(ad - bc)^2 \times n}{(a + b)(c + d)(a + c)(b + d)} = \frac{(750 - 50)^2 \times 100}{80 \times 20 \times 85 \times 15} = +24.0$$

If there is no association between A and B, the expected number of quadrats with A and B (that is, the expected value of a) is

$$\frac{[(a + b)(a + c)]}{n} = \frac{80 \times 85}{100} = \frac{6800}{100}$$

The other expected values are calculated by deducting the value of a from the margin totals. They appear on the above table in brackets. Because the calculated value of a (joint occurrences) is greater than expected, the association is positive.

87

In this method, the significance levels (2.4(a)) are always 3.84 at 5 per cent and 6.63 at 1 per cent (always one degree of freedom). The value in the above example (24.0) is greater than the 1 per cent level and the null hypothesis is rejected. The association index of +24.0 between A and B is declared 'highly significant'. So A and B really are associated and this probably means something biologically. The authors 'ecostat' computer program (2.9) can (naturally!) do this for you, but here is a good opportunity to write your own program.

Presenting data

If you do not have a computer, you may content yourself by comparing a few species with suspected associations, but to analyse your data fully, you need to take each species in turn, and compare it with all the others, again in turn. If you do this, present the results as illustrated below.

	Dandelion	Bent grass	Plantain
Daisy	+ 10.8	− 1.7	+ 0.8
Dandelion		− 2.0	+ 1.6
Bent grass			+ 3.5

Interpreting association indices

In the discussion of ecological niches (1.5) it was emphasised that the distribution of species is related to local environmental conditions because of adaptations possessed by them. Where they are positively associated, it usually means that they share similar habitat adaptations and occupy similar niches. Your physical measurements may help you to hypothesise what these might be. This method is greatly affected by quadrat size, reflecting the scale of the pattern. You could use your data from exercise 2.1 to investigate this.

In the above example, there is a highly significant positive association between Daisy and Dandelion. One hypothesis to explain this is that they both flourish best in places where the other plants do not grow so tall. Both species are low growing 'rosette' plants which are vulnerable to shading by taller plants. Another hypothesis is needed to explain why the plants of the other species do not grow so tall in certain parts of the site as they do in others. One likely possibility is that Dandelions and Daisies are more tolerant of trampling and grazing than are many taller species. Always remember that an association in one site may not exist in another.

6 Sampling Terrestrial Animals

6.1 PROBLEMS ENCOUNTERED WITH ANIMALS BUT NOT WITH PLANTS

These can be summarised as follows.

(a) Catchability

(i) Size

Depending on the habitat, either large or small animals can pose problems. If aquatic environments are studied the small size of some individuals not only makes them difficult to catch, but, once caught, they may be so small that they will remain unnoticed, and so will not be recorded. If, for example, traps such as the Longworth mammal trap (6.5), are used to capture terrestrial animals, some large mammals, such as the hedgehog, will be unable to fit into the trap and so will escape capture. It is important to be aware of the limitations of the apparatus and techniques used.

(ii) Speed of movement

Fast-moving animals can, obviously, run or swim away from the area being disturbed during sampling. On the other hand, slow-moving animals may appear in samples more frequently and so give a false impression of their density. For example, you think there are many more snails in a pond than there actually are, because you keep catching the same ones again and again. In this case, simply keeping all the snails you catch in one container until you have finished sampling the particular area will avoid the problem.

(iii) Camouflage

Because of their colours, some individuals blend in very well with their surroundings and so will be more difficult to see, which will lead to inaccurate recording.

(iv) Animal learning

This chiefly concerns mammals and is an important consideration when using the Longworth trap (6.5). These animals may soon learn that these traps contain food and offer comfortable shelter, and so they deliberately seek out the traps and keep returning to them. This leads to an over-estimate of population size. In a similar way, some animals may find the experience of being caught unpleasant, and consequently they become trap shy, resulting in an under-estimation of their numbers in the habitat.

(b) Migration

Emigration (animals leaving the population) and immigration (animals joining the population) will both have an effect on the population size at any one time.

(c) Birth-rate and death-rate

Birth-rate is the number of new animals entering the population per unit time. It is dependent on three factors.

(i) Fecundity: the number of eggs produced per female.
(ii) Generation time: the length of time between the egg-laying of the parent and when the offspring start egg-laying.
(iii) Sex ratio: the ratio of males to females.

Death-rate is dependent on the survivorship of the individuals from the moment they are born or hatched. The average life time of many inverte-brate species may be measured in weeks whilst that of mammals may be a matter of months or even years.

(d) Accurate identification

This is often difficult because of the sheer numbers of species. In fresh-water habitats in particular, many species need microscopic examination for a positive identification. Therefore, it often saves time if identification is restricted to Genus. Although this method is less precise, it is often easier for beginners.

Sampling animal populations on land usually takes place in woodland, scrub or pasture. The animals found in these habitats are soil organisms, litter invertebrates, airborne invertebrates, and mammals. Obviously, the type of animal you want to catch is an important consideration when deciding which equipment to use, as there is specific equipment available for capturing different animal types.

6.2 SAMPLING SOIL ORGANISMS

In chapter 2, the problems of counting daisies in a field were reduced to manageable proportions. Daisies, at least, grow very much in public, but many species of invertebrates live out of sight in the soil, and are difficult to sample unless they can somehow be persuaded to come out. The methods employed are called repulsion methods. The best known type of repulsion method is using smoke to drive wasps out of their nests. There are other more sophisticated methods, including the following.

(a) Earthworms
These are much more numerous than people think. Someone once said that in a field of cows there was more protein under the ground in the form of earthworms than on it in the form of steak. You may start to agree with this once you have carried out the following exercise.

Exercise 6.1. Estimating the number of earthworms

What you need
Each pair of students needs

(a) A quadrat, 0.5 x 0.5m.
(b) A plastic container with a screw cap, containing 2 litres of 0.5 per cent liquid detergent.
(c) Two jam jars.

A convenient means of transporting the detergent is in the undiluted form, making use of a supply of water (such as a stream) nearby. Some containers are taken empty, and are marked to indicate the level to which they must be filled so that they contain two litres. Small bottles containing enough undiluted detergent to give a 30 per cent final concentration when added to the water in these containers (taken from the stream etc.) are also taken. This reduces the transport difficulties, and enough small bottles of methanol can be taken to replenish the students' supply and enable them to repeat the procedure several times. For example, 10cm^3 liquid detergent made up to two litres gives a 0.5 per cent working strength.

Method
The area to be sampled must be defined, either as a field bounded by hedges, or marked out with tapes (perhaps as a grid or transect as part of

a wider study (exercises 2.1, 8.1, 8.2 and 8.3), depending on how you intend to sample). Ideally, you should sample two areas (for example, rough grassland and a cultivated field) and compare them.

The size of the area to be sampled must be determined either by direct measurement, or, in the case of a large field, by consulting a large scale Ordnance Survey map (1 : 25 000 or larger) which will have field boundaries marked. (When you have measured it, do not forget to multiply by 25 000!).

Next a series of random quadrats are taken (2.2). Each time the quadrat is laid down, two litres of methanol solution is poured over the area enclosed by the quadrat frame. After a few minutes the earthworms will start to appear on the surface of the ground. They must be quickly washed in water (to remove the methanol) and then collected in the other container. The population can be estimated as in the following example.

Area of quadrat = 0.5×0.5 m = 0.25 m^2
Area of field   = 200 m $\times 100$ m = $20\,000$ m^2

Quadrat	Number of worms
1	15
2	10
3	9
4	20
5	17
6	12

Total number of worms = 83

Average number of worms per quadrat = $\dfrac{83}{6}$ = $13.8 = 14$ per 0.25 m^2

Therefore, in 1 m^2 there is an average of $14 \times 4 = 56$ earthworms so the density
= 56 per m^2

and in a field of area $20\,000$ m^2 there will be $20\,000 \times 56 = 1\,120\,000$ earthworms

In your write up
1. Present your calculation of density and total population based on the whole class's results for both habitats, perhaps using the Mann–Whitney test (2.6).
2. If you have data on vegetation and habitat factors, discuss the ecological significance of your results in relation to them. Make hypotheses in terms of the biology of earthworms and their behaviour which may explain the results. For example, earthworms feed on humus in soil,

92

and soils with a high humus content usually have a high concentration of these animals. Earthworms are often less abundant in waterlogged soil because it lacks oxygen. There may be many more hypotheses that you can think of yourself.

3. Weigh ten randomly selected earthworms and determine their average mass. From this, you can calculate the mass of earthworms per 1 m^2 of habitat, and then work out the total mass of earthworms in the whole field.

4. You may have noticed that there are several species of earthworms. Try to identify them using reference books. If you decide to do a project based on this exercise, you could sample the various species to obtain densities for each.

(b) Soil arthropods
These dislike warmth and dryness

Exercise 6.2. Extracting soil arthropods using a Tüllgren funnel

What you need
(a) A Tüllgren funnel (figure 6.1)
(b) The soil sample.
(c) 30 per cent methanol (optional).

Method
A sample of soil is placed on the sieve and a lamp is placed a little way above the soil. The sieve rests on and opens into a funnel which opens into a collecting vessel which may or may not contain methanol to kill the animals as soon as they come into contact with it. It is best to have a pre-servative of some sort in the collecting vessel if you are going to leave the apparatus unattended for a period of time, as any arthropods collected could very easily crawl away leaving you with no sample. The apparatus should be left for at least two days if possible, to make sure the majority of the arthropods have emerged.

The funnel works on the principle that as the soil gets warmer and drier the arthropods move further and further down into the soil to get away from the source, until they finally reach the sieve and fall through the holes into the collecting vessel.

In your write up
The results that you obtain will be qualitative, not quantitative, so you cannot make any estimates of population density as you could in exercise

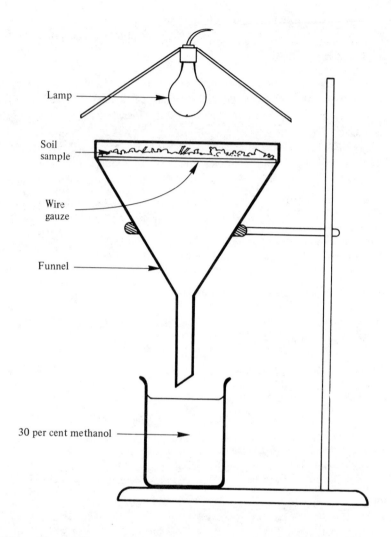

Lamp

Soil
sample

Wire
gauze

Funnel

30 per cent methanol

Figure 6.1 *A Tüllgren funnel (after figure 90 in T. King's,* Ecology, *published by Nelson)*

6.1. However, you could take several soil samples from different habitats and identify, using keys, any individuals you find. You could then compare the species found in different habitats, such as pastureland and woodland. Try and think of reasons why different species are only found in one of the habitats.

(c) Soil nematodes
This technique works on the principle that nematodes are heavier than water and therefore sink.

Exercise 6.3. Extracting soil nematodes using a Baermann funnel

What you need
(a) A Baermann funnel (figure 6.2).
(b) The soil sample.

Method
The soil sample is placed in a gauze bag and tied. This bag is then placed in the water in the funnel. The nematodes move from the soil through the gauze and into the water, and sink to the bottom. After about 10 hours the bottom 2 cm of water can be run off into the collecting vessel. The procedure can be speeded up slightly by gently heating the water by means of an overhead lamp. However, care must be taken to avoid overheating, because at temperatures of $30°C$ and above nematodes become paralysed and so are unable to swim. Obviously, this would affect the results.

In your write up
Again the results you obtain will be qualitative, so identify as many of the nematodes as possible, using a key.

6.3 SAMPLING LITTER INVERTEBRATES

(a) Quadrats
The simplest way to sample litter invertebrates is to use a quadrat. This can be carried out in conjunction with the collection of plant quadrat data, but as this is 'destructive sampling', make sure you do it last! The procedure is best carried out in woodland in autumn. It is particularly interesting to compare two types of woodland (exercise 8.1 and section 9.6).

Exercise 6.4. Sampling litter invertebrates using a quadrat

What you need
(a) A 0.5×0.5 m or a 1×1 m quadrat frame.
(b) A white tray.
(c) A trowel.

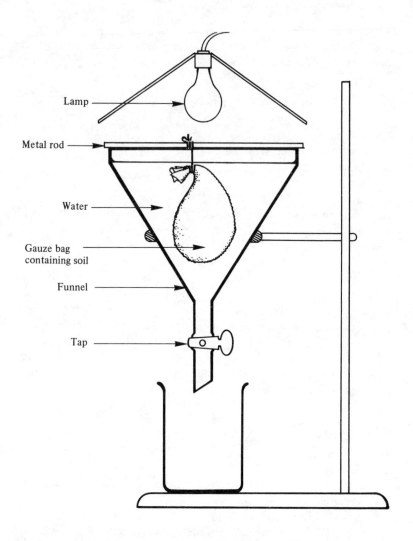

Figure 6.2 *A Baermann funnel (after figure 90 in T. King's,* Ecology, *published by Nelson)*

(d) A sieve.
(e) A pooter.
(f) A handlens.
(g) Specimen tubes.
(h) A record sheet.

Method

Prepare a suitable record sheet (2.8). Place a quadrat frame either randomly or regularly (2.2 and 2.3). Remove the top layer of debris. Next, using a trowel, remove the top 2 cm of litter and place it into a sieve which empties into a tray. By careful examination of the contents of the sieve and tray using a pooter (figure 6.3) and handlens, you should be able to identify many species (or Genera) of litter animals using a key.

By sucking up the animals into the pooter you can examine them, with no danger of them running away. However, it is too small to pick up large invertebrates such as beetles and woodlice, which must be dealt with using forceps. The specimen tube attached to the pooter can easily be replaced with another one, so samples once picked up do not have to be touched again. The animals can then be examined closely with a handlens.

These data can then be used for calculating diversity indices (5.6).

In your write up

1. If you have sampled two habitats try to think of reasons why the types of species are different, or why one species is very numerous in one habitat but rare in another.

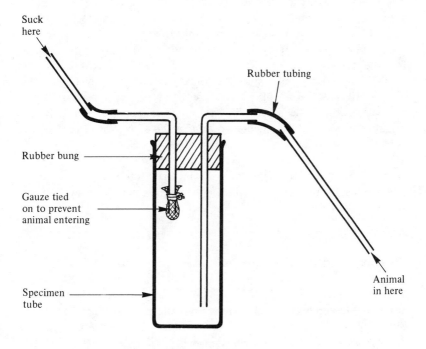

Figure 6.3 *A pooter*

2. Find out which of the species are herbivores and which are carnivores. Is there a difference in the total numbers of each? If there is, why do you think this is so?

(b) Pitfall traps

These traps are cheap and easy to obtain, because any old clean tin can, jam jar or drinks container can be used. The trap is laid by digging a hole in the ground, deep enough to take the whole tin or jar. It is important that the mouth of the container is level with the ground so that the animals are unaware of the trap, and drop into it rather than walk around it.

The traps must be protected from rainfall, otherwise they will fill with water. A flat stone supported by smaller stones or twigs will serve as a roof. However, if there is a shortage of suitable flat stones in the area, an ordinary tile or a plastic petri dish lid, weighed down with stones or soil from the hole, would do just as well. The roof may also serve to protect the trap from larger vertebrates (figure 6.4).

There is one disadvantage with this method of trapping. That is, once your animals have fallen into the trap they, hopefully, cannot get out — why might this be a problem? In any animal population there are carnivores and herbivores. When both are present in the bottom of a pitfall trap the carnivores will eat the herbivores, leaving you with 100 per cent carnivores in your sample. This gives a false impression of the population of animals as a whole. This problem can be overcome by placing 30 per cent methanol in the bottom of the trap, thus killing the animals the instant they fall in. This method of killing the animals is satisfactory if you just want to examine how many different species are present in the habitat. However, if you want to estimate the population of, for example, ground beetles, then the last thing you want to do is to kill them, so population estimates must be carried out at a separate time from any general survey of species present.

Laying out the traps at a chosen site

This method can be used to compare two sites, using a regular grid (2.3) or along a transect (8.5 and 8.6), as an exercise on its own, but it is more interesting to combine it with a study involving plants, other animals and environmental factors (exercises 8.1, 8.2 and 8.3). Figure 6.4 illustrates the organisation of a grid of traps. Remember that the Mann–Whitney test should not be used with regular sampling.

(a) *A correctly set up pitfall trap*

Roof

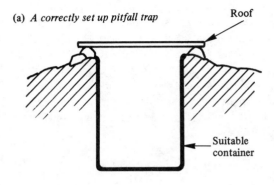

Suitable
container

Faults:

(b) *Trap not sunk in deep enough*

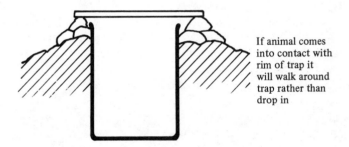

If animal comes
into contact with
rim of trap it
will walk around
trap rather than
drop in

(c) *Trap liable to flooding*

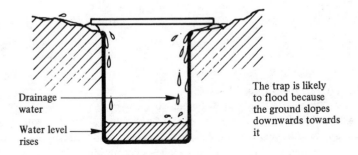

Drainage
water

Water level
rises

The trap is likely
to flood because
the ground slopes
downwards towards
it

Figure 6.4 *Setting up a pitfall trap correctly*

Examining your catch

Assuming that you have chosen not to have a preservative in your trap, your catch will be alive and running around, and this presents a problem – how do you pick them up and examine them closely? Firstly, the catch must be transferred from the trap to a specimen tube (labelled with the same grid reference, or transect line number). This is best achieved by using a filter funnel with a short stem placed on top of the specimen tube. The contents of the trap are tipped into it and the lid is then placed on the specimen tube. Use a pooter (figure 6.5) to pick up individual specimens if necessary (exercise 6.4).

Exercise 6.5. Sampling litter animals using pitfall traps

What you need
(a) Measuring tapes.
(b) Suitable tins, such as 450 g baked bean tins.
(c) Tiles or flat stones.
(d) Small stones for supporting the tiles.
(e) The bait (if you decide to use it).
(f) Waterproof pens.
(g) A pooter.
(h) Specimen tubes.
(i) A handlens.
(j) Forceps.

Method

Set up two grids (2.3, figure 6.4) in contrasting habitats, or contrasting parts of the same habitat. If vegetation sampling is not associated with the exercise, brief notes should be made about the distribution of vegetation, bare ground, footpaths and trees. When the grid (or transect) is laid out, each pair of students should set up at least one trap at 1 metre intervals to cover the whole transect or grid, and label them by writing on the tile with a waterproof pen. If you want to attract insects that usually visit flowers, the traps can be baited with jam or honey; or if you want to attract scavenger insects, animal faeces or old meat is useful. The traps can be left unbaited if you prefer.

The method of labelling must be agreed on before anyone starts; once the task is complete, all the tapes should be removed. The following day the traps should be emptied, as already described. The catch will then have to be taken back to the laboratory, where accurate identification can be carried out with the help of keys and reference books. Again, for simplicity it is probably advisable to restrict identification of most individuals to

Genus or Family. The results that everyone obtains should be entered on a communal record sheet, or blackboard table (table 6.1).

Table 6.1 A record sheet for a grid of pitfall traps

	A	B	C	D
1	Harvestman = 8 Spider = 2 Beetle = 6 Centipede = 1			Springtail = 2
2		Centipede = 2		
3	Beetle = 1	Spider = 1 Springtail = 1	Beetle = 1	
4	Springtail = 3 Harvestman = 2 Woodlice = 2		Spider = 1	Spider = 1 Snail = 1 Millipede = 1

In your write up

1. Write up the exercise, and in the results section you should present a copy of the complete table of results.
2. Which genera/species were present? If two grids were completed, each in a different habitat, emphasise differences in the species found in each.
3. In which parts of the grid were particular genera/species most abundant? Can this distribution be related to other factors? For example, perhaps some parts were more shady, more vegetated or near a path.
4. Which genera/species may have been under-recorded? Since some may have been eaten, it may help to find out which genera/species are herbivores and which are carnivores. Some genera/species may have been more likely to escape from the trap than others. Which?
5. Use the data to calculate diversity indices (5.6) and discuss the significance of these in ecological terms.

6.4 AIRBORNE INVERTEBRATES

(a) Capture using nets

Sweep nets are the most well-known types of net. This type of net is robust and can be used for 'bush beating'; this is when it is simply swept through long grass and undergrowth, and whatever falls into the net is then identified as far as possible and recorded. 'Tree beating' is a technique used where a large sheet is placed on the ground below a tree or bush and the branches are beaten using long poles. The animals from the tree will fall on to the sheet and can be identified. A Bignell beating tray can be used instead of a sheet; it is simply a collapsible frame with canvas stretched across it. This technique tends to collect the larval stages of flying inverte-brates, as the adults will simply fly away when disturbed. Hand-held nets can be used for catching the larger invertebrates such as butterflies, but this is time-consuming and not always rewarding. Butterfly nets are not sweep nets and should not be used for this purpose, because they are much lighter with a fine mesh.

(b) Sticky traps

These are simple to make from cardboard painted with a bait. The bait is usually based on black treacle which is boiled with twice as much brown sugar to make it stiff; some beer can also be added as this is particularly attractive for insects. The traps not only attract certain insects but some weak fliers such as aphids may be blown against the trap and get stuck on. The main problem with this method is that the catch is very difficult to identify unless the sticky substance is dissolved away from it.

(c) Water traps

These are the simplest of all traps. All that is needed is a shallow con-tainer of water with just a little washing up liquid added. This is widely recognised as the best way of trapping day-flying insects, and can be used quantitatively in the same way as pitfall traps are for ground-moving animals. If different coloured containers are used, different insects may be attracted.

(d) Light traps

Night-flying insects are attracted to light, and these traps exploit this fact. The simplest type consists of a white sheet laid against a bush or tree illuminated by a tungsten lamp. Moths, and other night-flying insects, are attracted by the light and settle on the sheet. Unless the sheet is coated with a thin layer of sticky substance (a large sheet of white card could be used instead) there is no way of keeping them there, so the only thing to do is to wait behind a bush with a net and catch them as they arrive.

Few people, except the more fanatical lepidopterist, would bother to do this, and it is much easier to use a commercially produced trap (figure 6.5). These moth traps have an ultra-violet tube which attracts the insects. They collide with the transparent vanes and fall into the box underneath

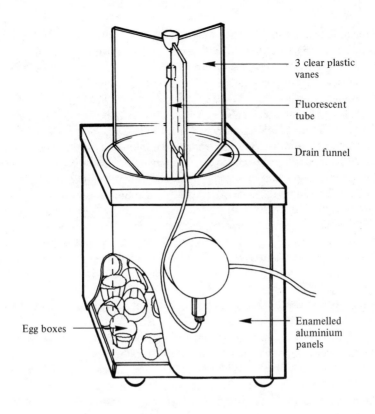

3 clear plastic vanes

Fluorescent tube

Drain funnel

Egg boxes

Enamelled aluminium panels

Figure 6.5 *A moth trap*

which is filled with egg boxes for them to settle in, from which they rarely can escape. These traps can be left on all night and examined in the morning. Although they are usually used with mains electricity, they can also operate from a car battery in field situations. There a large number of moth species and identification is difficult, so a detailed exercise is not really appropriate at this stage unless a great deal is already known about moths. The one exception is that you may be able to find out the proportion of moths in the area that are melanic without identifying the species. In industrial areas, many moth species occur in darker (melanic) forms than in the country. Melanism provides camouflage in such places, as moths settle on dirty buildings and trees; it is an advantage in that it protects them from predators (birds). As you go from a city into the country, the percentage of melanics in the population is reduced. You could find the ratio of melanic/non-melanic in different places and compare them. An added intriguing point is that some rural places (for example, the Shetland Island of Unst) have a high incidence of melanism, for reasons which are not fully understood. *The Natural History of Shetland* by Berry and Johnstone (Collins) in the 'New Naturalist' series has a useful section on non-industrial melanism.

Another possible exercise is to have a moth trap working throughout the year, or for a few months at least. During this time changes in the population of moths could be recorded from week to week — something which is impressive and related to weather and the flowering of the plants on which they feed. There are sometimes night-flying insects even in winter. The caddis fly, whose larvae is to be found in freshwater streams, is not a moth but a night-flying insect often caught in moth traps.

In project work an added sophistication might be to use the trap at different heights above the ground (for example, at ground level and on a flat roof) to investigate any differences in the distribution of flying insects.

6.5 SAMPLING MAMMALS

The Longworth mammal trap and the cheaper 'trip-trap' are two commercially available mammal traps which operate on the same principle. Although the trip-trap is less durable, it has the advantage of being transparent, so allowing you to see what you have caught without opening the trap. The following description applies mainly to the Longworth mammal trap.

The trap (figure 6.6) is made up of two parts: a tunnel which has a trap door at one end operated by a trip wire at the opposite end, and a nestbox into which the tunnel leads. The nestbox is arranged, when correctly set up, in such a way that it tilts towards the tunnel, and this is important because it ensures that the nestbox does not become flooded with urine which would foul the nesting material.

The trap is usually left in position for at least two days (but preferably longer) before it is actually set. The door can be fixed open during this time by pulling out the wire locking device until it clicks into place. To check if you have fixed it properly, simply press down on the trip wire — if the trap door shuts, try again. The trip-trap does not have this facility and so the plastic trip wire has to be removed or the door must be firmly fixed open, so that no animal is trapped.

During these few days prior to setting, the trap should have nesting material in it, and a trail of food leading into the tunnel. This will attract small mammals and encourage them to go into the trap (they will be able to leave the trap as the door will not shut behind them).

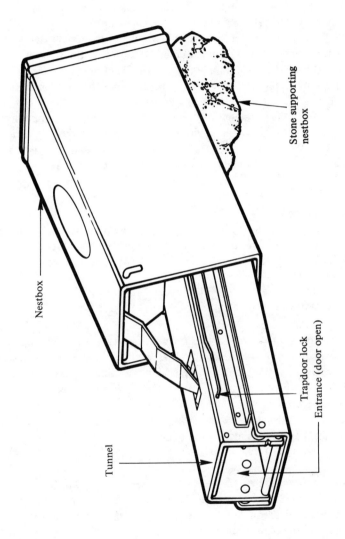

Nestbox

Stone supporting nestbox

Trapdoor lock

Entrance (door open)

Tunnel

Figure 6.6 *A Longworth mammal trap*

The trap is shiny aluminium so it must be camouflaged with leaves, twigs, etc., so that it is as inconspicuous as possible both to animals and to passers-by. The reason for setting up the trap and leaving it down for a few days without setting the trap door is so that the animals habituate to it (get used to it) and are, as far as possible, no longer frightened away by the strange object but are attracted by the food and nestbox.

Once the animals have the opportunity to habituate, each trap is set, and the next time a small mammal enters and touches the trip wire, the door closes behind it, so trapping it inside. Small animals have a high metabolic rate and it is important to supply enough food for at least twenty-four hours or the animal may die.

Even one or two traps are likely to yield interesting results, but at least twenty five are necessary to estimate population sizes (6.6). In serious ecological research, over 100 or even 200 might be used. All the traps available should be laid either at random (2.2) or on a grid (2.3).

(a) Grid trapping

Traps are placed 3 m apart, at the points of intersection of a grid, labelled with a suitable reference number and all facing the same direction. This can be included as part of an isonome study (8.12) or be done as a separate study.

(b) Line trapping

A line is laid down through the chosen area — again a transect line previously used to sample plant communities would be ideal but is not essential; the traps are set at regular intervals along this line. Figure 6.7 shows a line 70 m long. Five traps are set every 5 m in a rough circle, 1 m away from the measured point on the main line. Of course, shorter lines can be used, and the intervals between groups of traps can be altered to suit personal needs and the availability of equipment.

Examining your traps

The next day it is important to avoid touching the captured wild animal which may carry disease and is likely to bite. Therefore empty the contents of the nestbox into a suitable deep-sided container; a Perspex aquarium is ideal as it is light and easy to transport. When identification is complete, release the animal exactly where you caught it.

Mammal traps should be washed in a strong disinfectant after studies are complete, so that they can be stored safely.

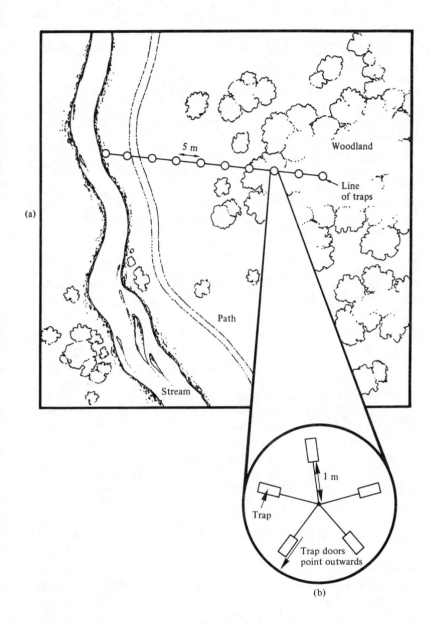

Figure 6.7 *Line trapping: (a) the site, (b) enlargement of each group of traps*

107

Exercise 6.6. Sampling small mammals using Longworth traps

What you need
(a) As many Longworth or trip traps as possible. The traps should be numbered.
(b) Bedding material, shredded paper or dry grass.
(c) The bait (seed mixtures as sold in pet shops for gerbils is suitable).
(d) Tapes.
(e) A supply of small stones for supporting the nestboxes of traps, if not available in habitat.
(f) Perspex aquaria.

Note
An NCC licence for capturing mammals must be possessed by your LEA if the traps you use do not have shrew holes, and you must place some meat inside for a captured shrew to eat.

Method
If possible, each pair of students should set up one or more traps either in any suitable place, or in a line, usually along a hedge or along a grid. They should be 3 m apart, if arranged regularly, set with the door open and left for at least two days. Then, late in the afternoon, they should be refurnished with bait, set to catch mammals and inspected the following morning.

In your write up
1. After describing where you positioned the traps, present the whole group's results in an appropriate manner and discuss them in relation to the distribution of vegetation and other factors. For example, are woodmice really more common in woods? If so, why?
2. In what ways would the use of 'break-back' traps (ordinary mouse traps) have been more or less satisfactory than using Longworth traps?

6.6 THE ESTIMATION OF POPULATION SIZE

The capture–recapture method
The procedure of capture, marking, release and recapture is an important one in animal ecology because it allows you not only to obtain an estimate of density, but in some methods also to make estimates of 'birth-rate' and 'death-rate' for the population being studied. If you capture animals, mark them and release them on more than one occasion; the population at any

one time will consist of some marked (*M*) and some unmarked (*U*) animals. Given this situation, you must know just two things to estimate the total population size

(a) The number of marked individuals alive (*M*).
(b) The proportion of the total population that is marked. That is the ratio

$$\frac{M}{M+U}$$

For example, if there were 200 marked animals and they made up one-third of the population, the total population must have been 600.

How can you get these two components? The second is easier to determine. You can estimate the proportion of the total population that is marked by drawing a random sample. You may assume that, if it is a random sample, it will contain the same proportion of marked animals as that in the whole population.

$$\frac{\text{No. of marked animals in sample}}{\text{Total caught in sample}} = \frac{\text{No. of marked animals in total population}}{\text{Total population size}}$$

(a) The Petersen Estimate or Lincoln Index
This is the simplest method used. There are two sampling periods

(a) Capture, mark, release.
(b) Capture, check for marks, release.

The time interval between taking the two samples must be short, because this method assures there is no recruitment of new individuals into the population between times 1 and 2.

Exercise 6.7. Estimating the population sizes of small litter animals using pitfall trapping

What you need
(a) Non-toxic waterproof paint and brushes.
(b) Specimen tubes, each numbered according to the trap in which the specimens were caught.
(c) Filter funnels minus necks.
(d) Pooters.
(e) Forceps.

Method

Mark all the individuals of the type whose population you wish to estimate, record the number n_1, and release them into the centre of the grid.
Next day, empty the traps and determine n_2 and m_2 (group results), and
calculate the estimated population ($\hat{N}$) using the formula below. There is
no reason why you could not deal with more than one animal type (for
example, ground beetles and centipedes) at once, but each will, of course,
need a separate calculation. You could repeat the procedure on successive
days to see how consistent your results are. To avoid confusion with the
previous day's marked animals, use paint of a different colour each day.

How to calculate the population size

Suppose that, using a grid of pitfall traps, you catch 120 ground beetles
(n_1). You mark them with paint and then release them into the centre of
the grid. Next day, the traps are re-examined and this time, 207 ground
beetles are caught (n_2). Of this second sample, 77 carry the distinctive
mark (m_2). So, from these figures, the proportion of the population
marked can be estimated

$$n_1 = 120 \quad n_2 = 207 \quad m_2 = 77$$

$$\hat{N} = \frac{n_1 \times n_2}{m_2} = \frac{120 \times 207}{77} = 322.5$$

where $\hat{N}$ is the total number of animals in the population.
Express your answer to the nearest whole number

$$\hat{N} = 322.5 = 323 \text{ beetles}$$

In your write up

1. Write up the exercise, including your calculation of population size.
2. You must assume that the population is closed. What does this mean,
   and what might be the effects if it is not, on the validity of your results?
3. All the animals have the same chance of being caught. Why might this
   not be the case?
4. The mark must be as small as possible, and be blue, red, brown, green
   or orange rather than white or yellow. Why?
5. The second sample must be a random sample of the population. Using
   pitfall traps, rather than catching animals which are wandering freely
   in the habitat, ensures that this condition is fulfilled. Why?

110

6. Only some types of paint are suitable for marking. Suggest two reasons why some paints might be unsuitable (apart from their colour), and how their use would make your estimate invalid.

(b) The Jolly–Seber method

This is a more complex method. It requires that the indivuduals are captured, marked, released, recaptured, checked for marks, marked again, released, recaptured, checked for marks, marked again, released and so on, many times.

This method, because it involves a larger sample taken over a longer period of time, yields more accurate results. However, the assumption that the population is closed can be relaxed.

Assumptions of this method

(a) Every animal, whether marked or unmarked, has the same probability of being captured.
(b) Every marked individual has the same probability of surviving from the ith to the $(i + 1)$th sample.
(c) Every animal caught in the ith sample has the same probability of being returned to the population.
(d) Marked animals do not lose their marks, and all marks are reported on recovery.
(e) Sampling time is negligible.

For the method to work, three or more samples are needed and you need to know when an individual animal was marked; for example, use dots of different coloured paint on beetles.

The equation for this method is

$$\hat{N}_i = \frac{M_i \times r_i}{n_i}$$

where M_i = total number of marked animals in population on day i
r_i = total number of animals recaptured on day i in sample
n_i = total number of animals in sample on day i.

Between marking and recapturing, emigration, immigration, birth-rate and death-rate are allowed for. So, you need to estimate the number of marked animals present in the population (M_i). For example: for a sampling session over five days

Table 6.2 Jolly–Seber method (reproduced by permission of Biometrika Trustees)

Day	Total captured (n_j)	Total released (a_j)	Day last captured (j)			
1	54	54				
2	146	143	10			
3	169	164	3	34		
4	209	202	5	18	33	
5	220	214	2	8	13	30
			$R_i = \sum\limits_{j=1}^{5} 20$	60	46	30
			R_1	R_2	R_3	R_4

where R_i = total number of captures for each of these days. So $R_1 = 20$ recaptured that were marked on day 1.

You are only concerned with the last mark. If you capture a marked animal, remark it with the last mark; that is, mark a day 1 individual again on day 2 as a day 2 individual, and so on.

Total number of animals recaptured on day i, bearing marks of day j or earlier

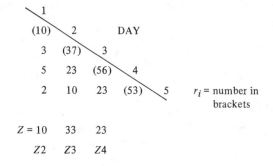

Calculate the total number of animals marked before time i that were not in the ith sample but were caught subsequently. So, they must have been present at time i.

Z is the number of animals that were not caught at t_2 but were subsequently caught, so they must have been present at t_2.

$$\hat{M}_i = \frac{a_i \times Z_i}{R_i} + r_i$$

112

For example

$$M_3 = \frac{164 \times 33}{46} + 37 = 154.6 = 155$$

$$\hat{N_i} = \frac{M_i \times n_i}{r_i} = \frac{155 \times 169}{56} = 707.9 = 708 \text{ animals in population}$$

At time i you mark a_i individuals and subsequently recapture R_i. You assume that the chances of recapture are the same, and all individuals have equal catchability.

$$\frac{Z_i}{\hat{M_i} - r_i} = \frac{R_i}{a_i}$$

$\hat{M_i}$ is the only unknown, so

$$\hat{M_i} = \frac{a_i \times Z_i}{R_i} + r_i$$

There are two groups of animals: ones that are captured and ones that are not. So you must assume that the recapture rate of those that you did not catch is the same as those that you did.

You can also estimate the probability of surviving from time i to time $i + 1$, which is given by the estimate of those in the population at time $i + 1$.

Survivorship

$$\phi_i = \frac{\hat{M_i} + 1}{\hat{M_i} - n_i + a_i}$$

Birth-rate

$i \rightarrow i + 1$
$$\hat{B_i} = \hat{N}_{i+1} - \phi_i(\hat{N_i} - n_i + a_i)$$
$i \rightarrow i + 1$

113

7 Sampling Aquatic Animals

7.1 SAMPLING AQUATIC ANIMALS

Just like terrestrial animals, aquatic life is very difficult to sample accurately because of its restricted catchability, the factors of migration, birth-rate and death-rate, and the complication of accurate identification. Accurate identification is particularly difficult because some species of genera need microscopic examination; for example, Mayfly are distinguished by the number or position of the gills. The aquatic environment includes both freshwater, marine and seashore habitats, most of which are suitable for ecological investigations.

The main use to which you will put the methods described in this chapter is in the comparison of two habitats, along the lines of exercise 8.1. You could compare, for example

(a) a fast-flowing and a slow-flowing stream (4.2 and 4.11)
(b) a eutrophic and an oligotrophic stream or pond (4.12)
(c) a polluted and an unpolluted stream or pond, or above and below a source of pollution in a single stream (4.2, 4.5, 4.6 and 4.7)
(d) a stream or pond on an acid moorland and one receiving water from a calcareous source (4.3, 4.6 and 4.9)
(e) rock pools on a seashore, or the same pool over a period of time (4.2, 4.3, 4.4, 4.6 and 4.8).

7.2 SAMPLING USING NETS

Probably the best way of sampling and capturing aquatic animals is by using nets. There are several different types of net, the most familiar type being the hand net. This is a versatile piece of equipment, as by using the same handle but changing the net for one of a different mesh size, different animal populations can be sampled.

(a) Plankton nets

The term 'plankton' is used to describe those pelagic organisms which are carried about by the movement of the water currents rather than by their own ability to swim. Some are only able to float on the surface of the water and are totally unable to swim, while others are weak swimmers but are carried large distances by the current. The planktonic animals are known as 'zooplankton' and the planktonic plants are termed 'phytoplankton'. While the majority of planktons are microscopic, there are some macroplankton which are visible to the naked eye.

There are many different plankton nets but they all consist of a long fine-mesh net with usually a 50–100 cm opening, held open by a loop of wire which is attached to a tow rope. At the end of the net there is a small container into which the collected plankton fall (figure 7.1).

The nets are often made of bolting silk or nylon, and it is woven in such a way that the mesh remains in place against the pressure of water. There are usually three mesh sizes: coarse (0.32 mm), medium (0.093 mm), and a fine net with 0.063 mm apertures. The mesh size used depends on what you want to sample; coarse mesh is more effective than fine mesh for catching large planktons, owing to the fact that it offers less resistance and a faster flow of water through it. Ideally, plankton nets should be towed behind a boat at a very slow speed, although samples can be taken by walking along the bank. The net can also be dropped into the water from the bank or from a platform and hauled up vertically.

Planktonic organisms are sensitive to temperature and so after sampling it is best to keep the sample in a thermos flask in some of the water until it can be examined. For accurate identification of your catch a microscope is needed, so examination should be carried out in the laboratory.

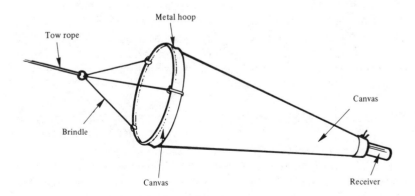

Figure 7.1 *A plankton net*

115

Exercise 7.1. A quantitative plankton study

What you need
(a) A plankton net.
(b) A flowmeter (optional).
(c) A thermos flask.
(d) A measuring cylinder (optional).
(e) Microscopes: one monocular and one binocular.
(f) A petri dish.
(g) A dropping pipette.
(h) Cavity microscope slides and cover slips.
(i) Graph paper.
(j) A haemocytometer.
(k) Plankton Identification Key.

Method
To measure the volume of water you are going to sample, a flowmeter can be attached to the net. A flowmeter has a rotating impeller which is moved by the water flow, while a counter records the number of revolutions. This can be placed in the opening of the net to measure the volume of water entering. Alternatively, a measuring cylinder could be filled with a volume of water, a sample from a stagnant pond is ideal and, by simply leaving the sample for some time to settle, a direct reading of gross quantity of plankton can be made from the scale on the side.

Having filtered a sample of plankton from a known quantity of water, the quantity present has to be estimated. The method of measurement depends on the type of study that you are carrying out and the size of mesh that was used.

Tip the contents of the thermos into a glass beaker; the large planktons are usually few in number and so can be picked out individually and placed in a petri dish. Accurate identification of each one need not be carried out, because it would be too time-consuming and difficult. However, an accurate count of the total number of individuals present must be carried out.

Smaller organisms may be so numerous that sub-samples have to be taken to make the numbers countable. The sub-samples can be spread out in a petri dish that is resting on some graph paper and examined, counted and recorded using a binocular microscope. The count can be made one square at a time. Smaller organisms could be counted using a haemocytometer, a microscope slide that is usually used for counting red blood cells.

During the course of counting you may notice that there is an outstanding number of one or two species; in this case it may prove rewarding to examine these more closely by placing them on a cavity slide, protecting them with a cover slip and using a Plankton Key to identify the species. You may find that the particular species is normally abundant in waters rich in particular minerals, and you could go on to take water samples to see if this is true in the case of the water that you are sampling.

Results
The results can be expressed in percentage form. For example

> 5 litres of water sampled
> 3571 individuals counted

> 1 litre = 1000 cm³
> 5 litres = 1000 × 5 = 5000 cm³
> No. of individuals per cm³ = $\dfrac{3571}{5000}$ = 0.7142

In your write up
1. Describe how you collected your sample and how you worked out the amount of plankton present.
2. Why and how may some plankton be lost from the sample?
3. If the population sampled does not remain constant during the sampling period, why?

(b) Nekton nets
Nekton are those pelagic animals which are more powerful swimmers than plankton and can travel from place to place independently of the water-flow.

The nets used to sample these animals are hand nets and framed nets — sometimes known as drag nets. Drag nets of fishing trawlers are large versions of the same thing.

Hand nets are easy to use, since all the operator has to do is to pull the net through the water at a pre-determined depth, then empty the contents of the net into a suitable container and examine the catch. A 'standard sweep' of a hand net can be defined as a sample taken at arm's length from the bank with the net held out as far as possible, the net then being pulled through the water for 1 metre or so. It is important to keep the mouth of the net below the surface of the water and to try to keep it at the same depth in the water so that mud etc. from the bottom of the stream or pool is not taken up too. Depending on the equipment that is available for a study of aquatic life, then the hand net can be used to sample the life in the mud or stones at the bottom.

Hand nets are often framed, but larger versions are available which can be dragged behind a small boat in much the same way as trawlers collect fish. Obviously, this method is only employed if the sea, a reservoir or a large lake is being sampled; in small stretches of streams, hand nets are usually adequate.

7.3 SEDIMENT SAMPLERS

Often it is quite adequate to use the hand net to dig about in the sediment at the bottom of a stretch of water. However, if quantitative sampling is required, other equipment can be used.

Provided that the water is shallow enough, then a framed net can be placed in a stream, with the mouth of the net facing upstream. Then, a quadrat can be placed at the mouth of the net (figure 7.2), and the soil within the quadrat is disturbed (by kicking?). The idea is that the current will carry the organisms downstream and they will be caught in the net. This does to some extent help to quantify the sampling area, and so some idea of the density of organisms per square metre can be obtained. A Surber Sampler is a more refined piece of equipment that has the quadrat frame attached to the framed net, and the quadrat has net sides that prevent the disturbed sediment from flowing past the entrance.

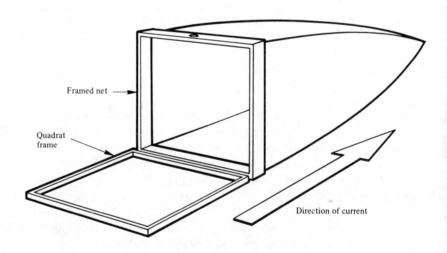

Framed net

Quadrat frame

Direction of current

Figure 7.2 *Framed net and quadrat*

7.4 WHERE TO CARRY OUT THE STUDY

Where you carry out a freshwater study depends on a variety of factors: how long you intend to study, how detailed it will be, or if you want to study the effects of pollution.

(a) Unpolluted water
A popular type of study is to investigate a pond, as many schools may have their own or have easy access to one. An advantage to studying a private pond is that equipment can be left out for a 24 hour period, with less chance of it being vandalised.

Exercise 7.2. Making a quantitative study of life in a pond, and taking physical and chemical measurements

What you need
(a) A hand net.
(b) A white tray.
(c) A wide-mouthed pipette.
(d) A handlens.
(e) A bucket.
(f) Specimen tubes.
(g) The identification key.
(h) Record sheets.
(i) Equipment for measuring: oxygen, temperature, light, pH etc. (chapter 4).
(j) Datamemory (optional).

Method
Working in pairs, or larger groups if necessary, in different regions of the pond, take 10 samples from the water using the hand net. The net may be pushed amongst the weeds, or in the mud at the bottom. Make sure that you sample both close to the bank and as far out as you can reach.

When you have a sample in your net, tip the contents out into the white tray and examine them carefully. It is advisable to use a shallow white tray; shallow so that the water is never too deep for you to see what is there, and white because everything will then show up clearly. A wide-mouthed pipette may be used to pick up individuals for close examination with the handlens. As each individual is examined and identified using a key, the numbers of each species caught must be counted and recorded on a record sheet which has been prepared beforehand. Although the term 'species' is being used here, it is not essential to identify to this level of accuracy. It is quite acceptable to restrict identification to genera, particularly if this is the first time that you have carried out a study of this type.

To avoid catching the same individuals twice, when you have dealt with each sample tip the contents of the tray into a bucket and take a further sample from the pond. Continue in this way until 10 samples (or as many as everyone has agreed on) have been taken. Any individuals that you cannot identify may be placed in a specimen tube and identified later.

Next take any physical and chemical measurements you want to and set up a datamemory to record some parameters over a 24 hour period (chapter 4 and section 2.10).

What to do with your data

If a microcomputer is available, the class results can be filed away. If there is no computer, then a master sheet of all data must be filled in by all members of the class. When this is complete, three things must be worked out.

1. The total number of individuals.
2. The total number of species (or genera).
3. The total number of individuals of each species (or genera).

Using these three numbers a diversity index can be worked out for the pond (5.6). The physical and chemical data will serve to illustrate the state of the water and any diurnal variations.

In your write up

1. Present your calculations, and comment on these results.
2. Can you explain why some individuals are present in larger numbers than others?
3. How do you relate the physical and chemical measurements to the ecology of the pond?
4. If you have made recordings over 24 hours, how can you explain any diurnal variations that may have occurred?

(b) Polluted water

If you want to study the effects of pollution, then obviously you must have access to a polluted stream or pond. You can, if you wish, study two streams or ponds, one of which is not polluted (as far as you know), but this is not necessary, and it is often advisable when studying a polluted stream to study both above and below the output of effluent so that all other variables remain as constant as possible.

To study the effects of effluent along a stream, a tape can be laid down along the bank and 10 m stretches marked out.

It is probably best to ensure that the effluent falls as near as possible to the middle of the tape, so that an equal number of samples are taken on both sides of the effluent; this makes data handling easier. Each pair of students can then study 10 m stretches, being careful to note exactly where they took their samples (The most dramatic effects are seen over several hundred metres with stations 500 m apart. This way you can also follow recovery if the effluent is not too bad.) Animals are the most useful indicators of water quality, so it is not necessary to measure physical and chemical factors.

Exercise 7.3. Studying a stretch of steam that has effluent running into it

What you need
(a) A hand net.
(b) A white tray.
(c) A handlens.
(d) A wide-mouthed pipette.
(e) Record sheets.
(f) A measuring tape.
(g) Wooden pegs.
(h) A mallet.
(i) Waterproof pens.
(j) Key to freshwater animals.
(k) Specimen tubes.

Method
Lay the measuring tape along the bank and try to allow the same distance above and below the outflow. Hammer a wooden peg at the beginning of the tape and then at 10 m intervals along it. Write the distance along the tape on each peg. If measurements of physical and chemical factors are not made at the same time, the pegs must be left in position so that exactly the same areas are sampled. The pegs must, of course, be removed when all necessary measurements have been taken.

In pairs, take 10 samples from the water using the hand net. Place each sample in the white tray and identify and record the number of each species found on a record sheet (table 7.1), before taking the next sample. There is no need to distinguish between samples on the record sheet, but if more than one 10 m stretch is sampled by the same people, a separate record sheet should be used.

Table 7.1 Example of a record sheet (see 2.8)

Distance along tape: 20–30 m

Genera	Number caught	Total
Gammarus	1 1	2
Ascellus	1111111	7
Daphnia	0	0
Hygrobates	1	1
Sialis	1 1	2
Corixa	111111111	9
Rhithrogena	111111111111111	15

The total numbers can then be written on a master sheet (table 7.2).

For each stretch of water studied, the number of individuals of each species must be totalled on a master sheet (table 7.2).

Results
When all the results have been obtained, histograms of some of the species can be drawn, either by hand or by using a computer.

Using the data, diversity indices for above and below the effluent (table 7.3) can be calculated (5.6).

From the above calculations it appears that there is a lower diversity of animal life after the effluent enters the stream. The Kothé system (7.6(a)) suggests that the stress suffered by the population below the outfall is minimal.

Table 7.2 Example of a master sheet

	Distance along tape (metres)						
				Area of effluent			
Genera	0–10	10–20	20–30	30–40	40–50	50–60	60–70
Gammarus	13	0	2	0	2	7	0
Ascellus	1	7	0	0	0	0	0
Daphnia	0	0	0	0	0	0	0
Hygrobates	0	1	6	5	7	9	5
Sialis	2	2	1	0	0	0	0
Corixa	3	9	4	0	20	21	23
Rhithrogena	20	35	2	0	0	0	0

Note: the tables serve as an illustration of record sheets and do not represent the data that were used for the calculations in the next section.

Table 7.3 Diversity indices

	Above outfall	*Below outfall*
Number of species (or genera)	38	32
Total no. of individuals	1250	1199
Simpson's diversity index	11.493	6.856

Kothé system (see 7.6) = 16 per cent.

Interpreting results

When interpreting the results of a survey a number of factors should be borne in mind.

(1) Natural changes will occur in the flora and fauna, which can be attributed to seasonal and diurnal variations, as will variation in physical factors. There may be some important event in the recent history of the stretch of water (such as a flood) that will explain some observations.

(2) The sampling technique used (for example, the mesh size of the net) may bias the results.

(3) The effect of pollution may encourage the growth of one or two species while others die. So, the survivors are there because they are tolerant and competition from non-tolerant species has been removed.

In your write up

In addition to plotting histograms (figure 8.3) and working out indices, there is a lot of work that can be done on the animals that are present. Perhaps some individuals appear for the first time after the outfall while others disappear at the same point. There must be a reason for this and part of the work will involve suggesting reasons for their distribution. The animals are good indicators of the degree of pollution present as well as what type of pollution it is.

Further investigations

Any study of pollution should be backed up with recordings of the physical and chemical factors in the stretch of water (chapter 4), to see if there are any correlations between certain chemicals such as nitrite and the abundance of, for example, algae. It should be noted that while chemical measurements give an indication of water quality at any one time, biological assessments may reflect conditions that have existed in the water over several weeks. Consider the example in chapter 11 (11.1 to 11.7).

7.5 SOME OF THE INDICATORS OF THE DEGREE OF POLLUTION

One of the best indicators of a rich supply of nutrients from sewage is sewage fungus, which grows on the surface of stones. The name 'sewage fungus' usually refers to one species of filamentous bacterium, *Sphaerolitus natans*, but some books refer to it as a complicated community of fungi, algae bacteria and protozoans.

If sewage fungus is present with a little oxygen, then it is likely that there will be a population of *Tubifex* worms. These live in the mud and can reach vast numbers (for example 420 000 per m^2) if there is no competition from other animals.

If there is no dissolved oxygen, then this is due to a high level of organic pollution which prevents the larger invertebrates surviving, as these need high oxygen levels to survive. If the water is shallow, then it is likely that large populations of *Culex ripens* will occur because of the lack of competition.

If there is a low level of organic pollution but a low oxygen level, then the sewage fungus will be sparse and *Tubifex* may be joined by the less tolerant red *Chironomids* – for example, *Chironomus riparius* (*C. thummi*) – which can survive in high concentrations of salts and ammonia.

Polluted but turbulent water can have a high level of oxygen. In this case the bed of the stream has a carpet of sewage fungus and there is a lot of *Tubifex* and red *Chironomids* in the mud. There may also be a small number of animals that are normally found in clean waters but are tolerant of the pollution, for example *Gammarus* and *Nais*.

If there is a sufficient level of nutrients in the water and a low level of solids, then algal growth increases and this increases the oxygen levels. The hog louse (*Ascellus aquaticus*) which is tolerant of low oxygen levels may appear. As the situation improves, molluscs such as *Sphaerium, Limnea pereger* and *Physa fontinalis* appear, together with leeches and some insects such as *Caddis* and *Sialis*.

Once the water becomes clean then Mayfly (for example, *Baetis*) that are tolerant of mild pollution may appear. However, Mayfly with three tail cerci and especially stonefly with two tail cerci are good indicators of clean water. Table 7.4 lists some of the commoner indicator species.

7.6 INDICES OF WATER QUALITY

Indices that are used for freshwater are based on chemical or biological observations, and there are some methods which link the two.

Table 7.4 Indicators of clean and polluted water

Animals found in water of very high quality only
Mayfly larvae – *Rhithrogena* (H)
Stonefly larvae – *Chloroperla* (C)
Freshwater shrimp – *Gammarus* (F)
Caddis larvae – *Rhycophila; Hydropsyche* (H)
Flatworms – *Crenobia; Dendrocoelom; Polycelis* (C)
Newts (C)
Frogs (C)
Toads (C)
Sponges (F)

Animals tolerating water of moderate quality
Jenkins spire shell – *Hydrobia jenkinsi* (F)
Limpet – *Ancylastrum* (H)
Ramshorn – *Planorbis* (H)
Orb shell – *Pisidium* (F)
Pea shell – *Sphaeridium* (F)
Damselfly larvae – *Coenagrion* (C)
Greater waterboatman – *Notonecta* (C)
Beetles – *Hydroporous; Gyrinus* (C)
Beetle larvae – *Dytiscus* (C)
Mites – *Hygrobates* (C)
Fish (C)

Animals tolerating water of poor quality (recovering from pollution)
Pond snails (but not those above)
Wandering snail – *Limnea pereger* (H)
Waterlouse – *Ascellus* (F)
Alderfly larvae – *Sialis* (F)
Lesser waterboatman – *Corixa* (F)
Water fleas – *Daphnia; Cyclops* (F)
Leeches – *Glossiphonia* (C)
Water cricket – *Velia* (C)
Pond measurer – *Hydrometridae* (C)
Pond skater – *Gerris* (C)

Note: the animals in this category are also found in better quality water, but because of competition, in smaller numbers.

Animals found in water of very poor quality (polluted)
Sludge worms – *Tubifex; Nais* (F)
Fly larvae – *Chironomus* (midge) (F)
             Chironomid (gel sac) (F)
             Chaoborus (phantom larvae) (F)
             Tipula (cranefly) (F)

(C) – carnivores, (H) – herbivores, (F) – filterfeeders.
Using the total number of carnivores and herbivores in your sample, you can work out the ratio of each.

It is only since the 1950s that invertebrates have been used as a guide to water quality. Research over the years has led to the development of a number of methods.

(a) Kothé system

This is particularly useful when comparing sites above and below an outfall of effluent, or it can be used to compare different sites. It assumes that the number of different species present is proportional to the degree of pollution. The greater the amount of stress due to pollution, the greater the decrease in the number of species present.

$$I = \frac{A - B}{A} \times 100$$

where I = index

A = number of species above outfall

B = number of species below outfall.

The higher the index, the greater the amount of stress.

(b) Trent biotic index

This method is probably the most useful for the inexperienced, as it involves minimal sampling. It uses invertebrates that maintain their position in water because they are attached to plants or stones. It also uses invertebrates that are affected by pollution and which are also easy to identify. It is useful in that it reduces the volume of data to a single figure that can be used for comparisons, and it is flexible in that it can be modified to take other factors, such as speed of water flow, into account.

The Trent biotic index (tables 7.5 and 7.6) works on a scale of 1 to 15, where 1 is very polluted and 15 is mildly polluted; it uses the large, easy-to-identify invertebrates and is weighted by the number of different groups of animals present. So, it considers diversity but it does not really involve abundance.

Table 7.6 is used as follows. If you have a sample which contains, for example two species of *Ephemera* nymph and nine other species then the biotic index is 7. If on the other hand, another sample has *Gammarus* in it, but none of the species above it on the table are present and 16 other species are present, then the biotic index is also 7.

This is another way of expressing diversity in a particularly useful application, used by Water Board inspectors. The higher the biotic index, then the more diverse is the water sampled. So, a high biotic index tells you the water is clean, while a lower one indicates some degree of pollution.

Table 7.5 Biotic index score sheet

1. Check list of animals below and calculate group scores.
2. Calculate total group score for each site.
3. Refer to table 7.6.

Invertebrate group	Group score
Flatworms	Score 1 for each species
Annelid worms (except *Nais*)	Score 1 for each species
Leeches	Score 1 for each species
Mollusca	Score 1 for each species
Crustacea	Score 1 for each species
Plecoptera (stonefly)	Score 1 for each species
Ephemeroptera (mayfly)	Score 1 for each genus
Baetis rhodani, a mayfly nymph	Score 1 if present
Trichoptera (caddis)	Score 1 for each family
Neuroptera (alderfly)	Score 1 for each species
Chironomid larvae	Score 1 for each family
Chironomus thummi, blood worm	Score 1 if present
Simulium larvae, black flies	Score 1 for each family
Other *Diptera* larvae	Score 1 for each species
Beetles and Beetle larvae	Score 1 for each species
Water mites	Score 1 for each species

Table 7.6 Trent biotic indices

Range: 0 = polluted to 15 = very clean.

Work down the group list until you reach the first type of animal in your sample. Read off the biotic index in the relevant group score column.

	Total number of species present for Biotic Index of									
	0-1	2-5	6-10	11-15	16-20	21-25	26-30	31-35	36-40	41-45
Plecoptera nymph >1 species	–	7	8	9	10	11	12	13	14	15
only one species	–	6	7	8	9	10	11	12	13	14
Ephemera nymph >1 species	–	6	7	8	9	10	11	12	13	14
only one species	–	5	6	7	8	9	10	11	12	13
Tricoptera larvae >1 species	–	5	6	7	8	9	10	11	12	13
only 1 species	4	4	5	6	7	8	9	10	11	12
Gammarus all above species absent	3	4	5	6	7	8	9	10	11	12
Ascellus all above species absent	2	3	4	5	6	7	8	9	10	11
Tubifex and/or Chironomid all above species absent	1	2	3	4	–	6	7	8	9	10
All above types absent	0	1	2	3	–	–	–	–	–	–
POLLUTED										

Reproduced with permission of the Severn Trent Water Authority.

8   Ecological Relationships

8.1   RELATING VEGETATION, ANIMALS AND ENVIRONMENTAL FACTORS

The various kinds of quadrat technique described in chapter 5 enable you to describe vegetation accurately and to begin to appreciate the relationships between plant species. Exercise 2.1 should have helped you realise that two superficially similar patches of grass differ in more ways than you might expect. Sampling animals (chapter 6) also reveals not only their amazing variety, but that their populations differ from habitat to habitat.

In the process, the idea (hypothesis) may have occurred to you that a particular species of beetle may be more abundant in one kind of vegetation than in the other. This could be confirmed (or otherwise) by recording the vegetation of both sites by quadrat, and simultaneously sampling the beetles by capture–recapture. You could then consider whether the differences between the sites with respect to the beetles and particular species of plants were significant. A question which would soon follow, perhaps, is "why does this beetle and certain plant species occur together in one place when they are all absent from a place nearby?" To explore such a question the approach needs to be broadened. To make an ecological comparison, not only must plant and animal populations be sampled, but also environmental factors, such as diurnal (daily) fluctuations in light, temperature and air humidity. Samples may be taken using various specialised equipment instead of quadrats and pitfall traps, but the process will use the same sampling grid. These non-biological data can be submitted to the same statistical tests of significance.

8.2   INTERPRETING ECOLOGICAL RELATIONSHIPS

These kind of relationships, based on field observations, even when analysed by statistical methods, do not usually prove anything, although they are important in getting the most out of your data. They must be interpreted with care.

Suppose, for example, with the help of the Mann–Whitney test, you show that field A has a significantly higher density of daisies than field B. If your sampling was satisfactory this would be a fact. Suppose you also find, using a similar sampling procedure, that field A has a significantly higher soil pH than field B — another fact. These two related facts invite an explanation, but you are soon in the realms of hypotheses, rather than facts. One attractive hypothesis is that daisies grow better at the pH of field A rather than that of field B. Such a hypothesis needs to be tested by experiment (chapter 11) and, whilst it is true that growth of daisies can be affected by soil pH, you might well find that the actual difference in pH between the two fields is not big enough to explain the difference in density of the daisies. Daisies require a higher level of soil nutrients than many wild species, and also flourish best with intensive grazing or frequent mowing. Field A may happen to be more cultivated than field B, and used more for grazing. Cultivation would perhaps involve fertiliser addition, which, together with grazing, might be the real explanation for the high density of daisies. Liming may well accompany the fertiliser application, thus explaining the difference in pH of the soil. Hence the relationship between daisies and soil pH would be an indisputable fact but you can (and must) merely hypothesise within a potentially complex situation when starting to look for an explanation.

The serpentine heaths of the Shetland Island of Unst provide a more exotic example. Daisies are comparatively rare, probably because of the low nutrient status of the soil, and their presence (along with certain other nutrient-loving species) in certain places usually implies some nutrient enrichment. This can often be explained by local concentrations of sea birds producing guano, or agriculture in the past (in the vicinity of ruined crofts and field walls). Daisies and other nutrient-loving species (such as the grass *Holcus lanatus* — Yorkshire Fog) are also common around old nineteenth century chromite quarries. Perhaps this enrichment can be explained by the dung of ponies used to carry ore (there is no evidence ponies were used, but it is a reasonable possibility) or by the workers' own toilet arrangements. The hypothesis that distribution of daisies in this habitat is influenced by plant nutrients can be submitted to experiment. Some hypotheses about the origin of local nutrient enrichment may need to be regarded as untestable (exercise 11.1(b)).

8.3 DECIDING WHICH ENVIRONMENTAL FACTORS TO MEASURE

A variety of factors that you could consider are included, along with techniques and background information, in chapter 3. Since you probably have no previous experience of the site and may be unlikely to revisit it, a certain hit-and-miss element will be inevitable in your choice of factors. You will include some which you regard as likely to be irrelevant, but some of these may lead to surprises. Lack of equipment or suitable techniques will restrict your choice. Sherlock Holmes might well have considered all possibilities, but he never had so many suspects. Limitations of time will mean that you must be more selective. Here are a few hints on making the necessary decisions.

(a) If possible, collect your vegetation and/or animal data, examine them and discuss them amongst your class before deciding which factors should be considered. Return to the site another day to collect the environmental data.
(b) Make a few preliminary measurements. A few soil pH determinations from both sites may turn out to be all more-or-less the same, suggesting a more detailed collection of this type of data might well not be worth the effort.
(c) Use the advice on specific habitats given in chapter 9.
(d) Consider any background information that you have on the site.

Exercise 8.1. Comparing two homogeneous habitats

This represents one approach to relating plants and animals to environmental factors. Two other approaches, transects and isonome studies, will be considered later in this chapter. This exercise is of the type discussed above, in which you select two sites, each of which is considered to be homogeneous (that is, the vegetation of each study area is apparently even throughout, like the pattern on wallpaper). If the vegetation tends to change from one part of the study area to another, this is not the appropriate approach. In practice, no vegetation is perfectly homogeneous, but, for example, two areas of pasture may be such that they can be considered so. A preliminary transect (exercise 8.2) may help to identify homogeneous patches of vegetation but this is not essential. You can compare any two sites but a few suggestions are made for terrestrial habitats (9.6), and, although the following exercise refers particularly to land habitats, the principles it involves apply in freshwater (7.1) and on the seashore (chapter 10).

What you must decide
(a) Where to place your grids, and the size of the grids (it will be easier to make both the same size, particularly if you intend to use the Mann-Whitney significance level table in appendix B).

(b) The type and size of quadrat and the number of quadrats to be recorded. If you work as a team with other students, you will be able to collect plenty of data in a limited time.

(c) The method of recording (percentage cover, Domin or point quadrat per unit area – chapter 5). Use point quadrats if you plan to calculate diversity indices (5.6). You must record at random (2.2(b)) to use Mann-Whitney.

(d) Which species to record. Prepare species list (2.8, notes following table 8.1).

(e) Whether to collect animal data, and, if so, the methods to use (chapter 6).

(f) Which environmental factors to measure. If possible, delay this decision until you have collected, examined and discussed your quadrat data. This work could be shared by a team.

What you need

(a) Suitable quadrat frames and recording sheets.

(b) Tapes, random number tables etc., for laying out a sampling grid.

(c) Appropriate equipment for any physical measurements you decide to make (chapter 3) and for estimating animal populations (chapter 6).

Method

The general approach is as in exercise 2.1. Lay out a sampling grid in each site. To ensure a reasonable degree of homogeneity, avoid including such things as the edges of fields. For each site, record the chosen number of quadrats (preferably the same for each), sample animals and make the same number of environmental recordings for each chosen factor.

Handling the data

(a) For each species of plant and animal and each environmental factor, calculate a mean value for each site and present it as a table as in table 2.3.

(b) Test each pair of datasets (for each species and environmental factor), comparing the two sites for significant difference using the Mann-Whitney test (2.6), and add 'NS', 'S*' or 'S**' to the table (as in table 2.3). These calculations need to be shared amongst the class to reduce the task to manageable proportions.

(c) Calculate diversity indices (if you wish, provided that you have used point quadrats) for each site (5.6) and add them to the table.

(d) You may wish to test some of the species of plants and animals for association (5.7).

In your write up

1. Summarise the main conclusions from the data; for example, "there was a significant difference in soil pH between the two fields. Long Meadow had a mean value of 5.4, whilst for Ten Acre it was 6.7."

2. Suggest hypotheses to explain the results; for example, "the differences in soil pH are probably related to the underlying rock, carboniferous limestone in the case of Ten Acre and sandstone for Long Meadow. This presumably explained the significantly higher cover of calcicolous species in Ten Acre." Chapter 3 includes background information for physical factors.

3. Select one hypothesis from your discussion and briefly explain how you would attempt to test it by experiment.

8.4 HETEROGENEITY IN VEGETATION: ZONATION, GRADIENTS AND MOSAICS

In many sites, the vegetation is clearly not homogeneous. This non-uniformity (heterogeneity) can take various forms. The most obvious may be zonation. If, for example, you ascend a Scottish mountain, the lower slopes near sea-level may have meadows or Oak forest. Above this might be a zone of Birch and Pine, but there will be tree-line at about 500 m (2000 m in the Alps). Above this will be a zone of heath (well-drained acid soil), peat bog (poorly drained acid soil) or grassland (neutral to alkaline soil, especially if grazed). The summit may only support mosses and lichens with small herbs, ferns and grasses in rock crevices. This zonation may well be further complicated by human influence — for example, much of the forest may be Forestry Commission plantation, natural forest may have been felled in the past and now be sheep pasture, and the heath may be managed as grouse moor — but the underlying pattern is related to altitude, giving rise to environmental gradients.

An environmental gradient is a gradual change with respect to a particular factor. There is, for example, a gradient of temperature related to altitude. There is a fall of approximately $1°C$ for each 150 m ascended, and the effective length of the growing season is reduced. At the same time, the amount of precipitation and wind exposure increases with altitude — two further environmental gradients. Such environmental gradients are reflected in gradients of vegetation and associated animal populations. A particularly clear example of zonation is evident on the seashore (chapter 10), where the gradients are related to the degree of exposure by the tide, which increases as you ascend the beach. Another example is shown in figure 8.2, involving a Cotswold escarpment. Here, the

133

degree of slope influences zonation, but the main factors involve human aspects. The pasture at the foot of the slope is affected by agriculture, whilst that at the top is managed as a golf course. The woodland on the slope has probably survived because the slope is too steep to plough, and the trees may, in any case, have been planted. Sand dune zonation (figure 9.1) represents a more complex situation (9.2).

Where the zonation is very clear cut, the best way to study it may be to recognise the zones and treat each as a homogeneous piece of vegetation, comparing them using the approach of exercise 8.1, using Mann–Whitney as a test of significance. Where there is a gradual transition with gradients of vegetation, implying environmental gradients, the best approach may be to use a transect (exercise 8.2) or an isonome study (exercise 8.3). The seashore is one such example. A woodland margin may provide another. In the Cotswold escarpment example (figures 8.2 and 8.3), between the wood and the upper pasture was a belt of scrub. It emerged gradually from the pasture, getting thicker and taller until it merged into the wood. At ground level, light intensity fell on approaching the wood – that is, there was a gradient of light intensity. Study of this involves not Mann–Whitney, but correlation coefficients (8.9).

A woodland floor often has patches of vegetation interspersed with areas of bare ground. There is no neat zonation readily described by a sketch map (like figure 8.2), but the vegetation represents a mosaic, related to gradients of light intensity caused by differing degrees of light and shade provided by the canopy of tree leaves. This can be studied by an isonome study (exercise 8.3), and, again correlation coefficients (8.9) provide the appropriate statistical approach.

8.5 TRANSECTS

A transect is a line (normally straight) along which observations are made in a systematic fashion. It is a form of regular sampling. It is used

(a) To describe zonation of vegetation and relate it to environmental factors. This application is illustrated by figures 8.2 and 8.3 and forms a familiar aspect of many elementary field courses. Its main value is as an introduction to ecology which requires no mathematical analysis. The results are often difficult to interpret and some people would consider the time better spent on a more specific problem, in which a more statistical approach is possible. You should regard it as a preliminary study which will help you to identify zones for further investigation for comparison along the lines of exercise 8.1, or gradients to be investigated by shorter transects with more frequent sampling, or isonome studies (exercise 8.3).

(b) To study known gradients of vegetation and animal populations and to relate them to gradients in environmental factors. Such an approach can provide detailed information about the transition from one zone to the next, where one zone merges into the next, rather than changes abruptly. It offers the possibility of statistical analysis by correlation coefficient (described below).

Once you have (subjectively) decided where to mark out your transect, all subsequent recordings should be made objectively, at regular intervals, using the methods described in chapters 3, 5 and 6.

8.6 TYPES OF TRANSECT

(1) A line transect is one in which plants or seashore invertebrates are recorded where they touch the tape. Record individuals touching the tape at frequent intervals (say every 1 cm or 10 cm), making a list covering from one end of the transect to the other. This method can be carried out quickly and is useful in a preliminary study, but it is crude and of very limited value from a statistical point of view.

(2) Belt transect: this term is commonly used for what was once known as a ladder transect. Originally, two parallel lines were marked out, and the 'belt' of vegetation between was described in its entirety. It is now more common to sample the 'belt' at intervals. Strictly speaking, therefore, a ladder transect is a belt transect sampled using regularly located quadrats in line between two points (figure 8.1). Quadrats are arranged such that one side of the quadrat frame lies alongside the tape (or string tied to marker pegs).

Exercise 8.2. Recording a transect

As a preliminary study (8.5(a)), you could use a transect anywhere where there is some zonation evident. The perhaps more satisfactory application (8.5(b)), the study of vegetational and environmental gradients, requires a more careful selection of the site. The seashore offers a particularly suitable situation (chapter 10), but, as mentioned above, a transition from woodland through scrub to open pasture may also be very appropriate. Other suggestions are made in chapter 9. There is no point in recording a transect unless you are aware that some kind of zonation exists.

136

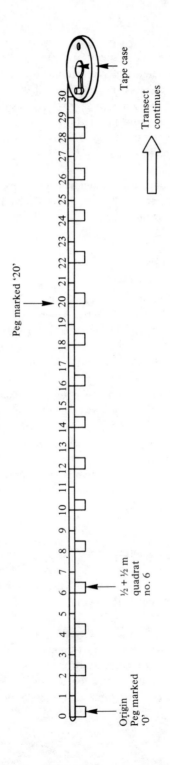

Figure 8.1 *Laying out a ladder type of belt transect*

Tape case

Transect continues

Peg marked '20'

½ + ½ m quadrat no. 6

Origin Peg marked '0'

What you must decide

(a) Where to lay out your transect, and how long it is to be. Where the zonation takes the form of parallel belts (figure 8.2) the transect should cross them roughly at right angles.

(b) The type of quadrat and method of recording (chapter 5). In most cases, 0.5 x 0.5 m recorded by percentage cover will be suitable. In some situations, point quadrat per unit area (5.5) may be appropriate. In a very long transect sampled infrequently (such as every 10 m up a mountain), a larger quadrat size of 1 x 1 m is recommended.

(c) How frequently to record. Figure 8.1 shows how the quadrat frames are placed at regular intervals. The distance apart depends on the length of the transect and the time available. The general principle is to sample as frequently as possible. Where the transect is very long, the whole class could work as a team (details of organisation are given below).

(d) Which species to record: prepare species list (2.8, notes following table 8.1).

(e) Whether to collect animal data, and, if so, the methods to use (chapter 6). You could combine a study of vegetation with, for example, earthworm density sampling (6.2(a)), pitfall trapping (6.3(b)) and Longworth mammal trapping (6.5).

(f) Which environmental factors to measure. If possible, delay this until you have collected, examined and discussed your quadrat data. The object is to measure factors which may vary (that is, form gradients) along the transect and be related to the pattern in vegetation. See the suggestions offered in chapter 9.

What you need

(a) Several 30 m tapes (depending on the length of the transect).

(b) A supply of wooden pegs, mallet and felt-tip pens to number the pegs.

(c) Recording sheets with species in standard order (see table 8.1) and notes to aid identification.

(d) Equipment to measure environmental factors. Refer to chapter 3.

(e) Equipment to sample motile animal populations (if you decide to consider them).

Method

(a) Prepare a recording sheet (2.8, table 8.1).

(b) Lay out the tape. Use several tapes end to end if necessary, and place numbered pegs at regular intervals. These can be left if you choose to record environmental factors on a separate occasion. They also help if

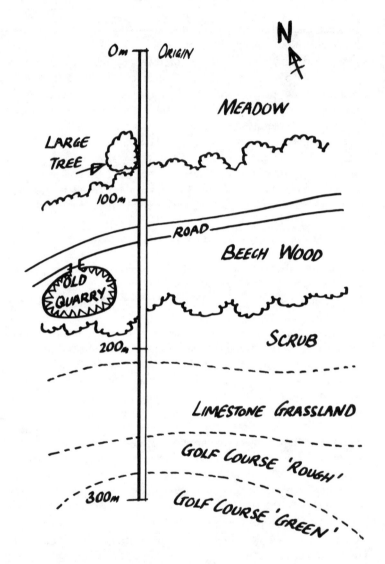

Figure 8.2 *Site sketch made showing zonation of a Cotswold escarpment. Under the sketch list: (i) the name of the site, (ii) its Ordnance Survey grid reference, (iii) the underlying rock type (in this case oolitic limestone) or surface deposit (for example, glacial drift, river gravel), and (iv) other information, such as north-facing or south-facing slope*

Table 8.1 Record sheet for a transect study

Species	Percentage cover for quadrat no.										
	2	22	42	62	82	102	122	142	162	182	202
F. sylvaticus	0	0	0	0	0	10	100	100	75	0	0
H. lanatus	30	35	20	10	0	0	0	0	0	0	10
Festuca spp.	40	30	35	5	15	0	0	0	0	0	20
Bromus erectus	0	0	0	0	0	0	0	0	0	0	40
P. intermedia	0	0	0	0	0	0	0	0	0	0	5
M. perennis	0	0	0	0	20	70	10	50	30	5	0
P. purum	0	0	0	0	0	0	0	0	0	5	15

Notes
(1) Only part of the transect and 7 of the 65 species recorded have been presented.
(2) In the species list, some generic names have been abbreviated (for example, *Bromus erectus = B. erectus*).
(3) In '*Festuca* spp.', spp. means 'species' (plural). Two similar species, *F. ovina* (Sheep's Fescue) and *F. rubra* (Red Fescue), were present at the site but the recorders were unable to differentiate them reliably in every quadrat. Recording them separately (one keen student claimed she could tell them apart but it was all 'wiry grass' to the rest) would have produced impressive but spurious data.
(4) The layers in a woodland (5.2(c)) give much more than 100 per cent cover. A value of 75 per cent (see quadrat 162) for *Fagus sylvaticus* means that 75 per cent of the quadrat was overhung by Beech foliage.
(5) These data, combined with those of the rest of the class, are presented for the complete 300 m transect, as in figure 8.3, with some physical data.

you do not have enough tapes, and have to keep moving them along. Pegs cannot, of course, be inserted into a rocky shore.

(c) Record quadrats at regular intervals, and collect environmental data (and perhaps set pitfall traps etc.) to correspond with each quadrat. If you have decided to record environmental data on a separate occasion, they should be collected from exactly the same places in which the original quadrats were placed, and the quadrat data should correspond exactly to the original data. You may record a complete transect yourself (if it is short), or as part of a team (see below).

It is all too easy to get your data mixed up. Number your pegs when you position them and use the record sheet strictly for all data. Avoid having some of it in separate notebooks to transfer to the sheet later.

Presenting the data
Each species and physical factor should be represented as a histogram (or kite diagram – figure 10.2) in which distance along the transect is on the

horizontal axis (see figure 8.3). To aid comparison, as many separate histograms as possible should be drawn on the same sheet of graph paper, with a common horizontal axis, even if the sheet is very long (perhaps several fixed together). This process of collation can be greatly speeded up using a microcomputer (2.9). The sooner the data is summarised on paper or VDU screen for discussion of the biological significance, the better.

In your write up
1. Draw a simple site map (figure 8.2) and summarise any relevant background information (land use, history, geology, etc.).
2. Briefly describe the vegetation along the transect, mentioning dominant species in the various parts of it. Emphasise other species which are restricted to certain parts and those which seem to be evenly distributed along it. Where a community seems to represent a transition between two other communities, could this imply succession (9.4)? *Note*: succession is not an important feature of the seashore.
3. Relate any animal data that you have collected to the vegetation.
4. Which physical factors show environmental gradients along the transect? In some cases, the gradient may only exist in parts of the transect, and, if so, where? In the example (figure 8.3), there is a relationship between light intensity and some aspects of the zonation between 60 and 220 metres. Soil pH, on the other hand, varies little (between 6 and 7.5). This supports the belief (suggested by the local geological map) that all the soil along the transect is derived from the underlying limestone rock. No relationship is evident between soil pH and vegetation. You will get the most out of your data if you read sections 8.7, 8.8, 8.10 and 8.11, and even more with the help of section 8.9 as well, before proceeding. Identify as many correlations as possible. Discuss the relationship between environmental factors and plant and animal species using background information and points for discussion in the appropriate sections of chapter 3 dealing with the methods.
5. To what extent can the pattern in vegetation be explained by human activity and related biotic factors, past and present (1.7 and 9.2)?

Working as a team on one transect
If the transect is very long, the whole class may combine and collate their data later. This can be very useful but confusion will result unless there is good organisation. The following procedure ensures that each individual examines points along the full length of the transect. Reference is made to a particular example and you will need to modify the procedure according to the length of your transect, the number of students involved, the physical factors you decide to measure and local circumstances.

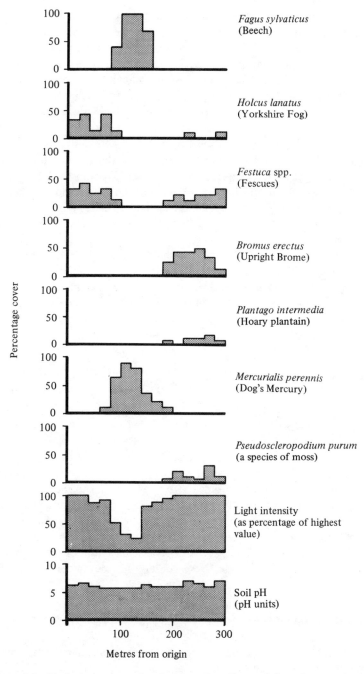

Figure 8.3 *Displaying transect data by histograms. Some of these data are given in table 8.1*

If there are twenty students working in pairs and the transect is 300 m long, divide it into 20 m sections using marker pegs, labelling each peg — the first, the origin, 0, the next 20, and so on (see figure 8.1). The aim is to record a 0.5 × 0.5 m quadrat every second metre by percentage cover. It is essential that the whole group agree on a numbering system. Locate a quadrat with one side along the transect line, extending from the origin marker peg to 0.5 m on the tape which should be marking the transect. Call this quadrat 0. Quadrat 2 covers 2 to 2.5 m, and so on, and quadrat 20 covers from 20 m (with a labelled peg) to 20.5 m. According to this system, the number of each quadrat relates directly to a position on the ground. Each pair of students records one quadrat from each 20 m section. The first pair records quadrats 0, 20, 30 etc., and the second quadrats 2, 22, 32 etc. Every quadrat should be numbered carefully on the record sheet. If there are 13 pairs of students sampling every 2 m, then the sections will be each 13 × 2 = 26 m.

Measurement of physical factors should be approached differently. Taking twenty students measuring five factors, they could work in fours, each team taking one of the factors, and organising themselves after having read the appropriate parts of chapter 3. They may not record every second metre, as with the vegetation, but they must be able to relate their factor directly to the quadrats. Aim to have at least one physical measurement for each section, but preferably one corresponding to each quadrat.

Collating the team's quadrat data
This should be done as soon as possible on returning from the field — quickly and accurately, lest the objective of the exercise be lost amongst a mass of mud-stained pieces of paper. A new copy of the record sheet should be prepared for each section of the transect and clearly headed (for example 0-20 metres) and the columns labelled for each quadrat in the section (0, 2, 4, 6, 8 etc. to 18). These sheets are circulated amongst the class, and each pair of students transfers its data to the appropriate columns. When each sheet is complete, average values are calculated for each species. A master sheet is then prepared in which each column contains average values for each section of the transect. If you have made more than one physical measurement for each section, calculate section averages in a similar fashion. Now consult 'In your write up' section (above).

8.7 CORRELATION

If you have not carried out exercise 8.2, you should nevertheless study figure 8.3 and read the 'In your write up' section, as the example is used in the following discussion. We observed that the plant species Dog's Mercury became more abundant on entering the wood, and that this was associated with a decreasing light intensity. There was a mathematical relationship, a correlation, between the two sets of data. Since it was a case of the more light there was, the less Dog's Mercury, the correlation was a negative one. There appears to be a positive correlation, on the other hand, between light intensity and the distribution of the common grass *Festuca* spp.

8.8 SCATTER DIAGRAMS

Examining histograms is a perfectly satisfactory way of identifying possible correlations, including whether they are positive or negative. A scatter diagram is a means by which you could pursue the matter further if you wish to get more out of your data. To compare two sets of data, for example, the cover values for Dog's Mercury against light intensity from transect data (figure 8.3), draw a graph with light on one axis and Dog's Mercury on the other, and plot a point for each quadrat (or the mean value for each section) on these axes. The scale you use should be such as to spread each dataset along most of the relevant axis, and need not start at zero. Soil pH, for example, could be plotted between 4 and 8. Inspection of the original histograms (figure 8.3) had shown that the gradient of light intensity was restricted to the 60–219 m part of the transect. In this example, the values plotted are mean values for each 20 m section. It would, at this stage, be a good idea to produce a new set of histograms confined to this part of the transect using individual quadrats instead of section means, and also individual light intensity values for each quadrat. Alternatively, you could return to this part of the site and do a more detailed transect, or an isonome study (exercise 8.3). As has already been pointed out, a transect is particularly useful in studying this type of gradient. Elsewhere, light values were all the same (100 per cent of the maximum value), and Dog's Mercury was completely absent. Data for the 60–219 m part of the transect are presented in figure 8.4 and plotted as a scatter diagram (using section means for simplicity). A line of 'best fit' has been added, and it can be seen that the points lie scattered closely about it (perhaps the 'best fit' should be slightly curved?), indicating a negative correlation because it slopes downwards from left to right – the less light, the more Dog's Mercury.

Figure 8.5 will help you to interpret your own scatter diagrams. In a 'perfect' correlation (a and b), the points lie on a line (straight in this case) The line may not be straight (c). In very dense woodland, Dog's Mercury

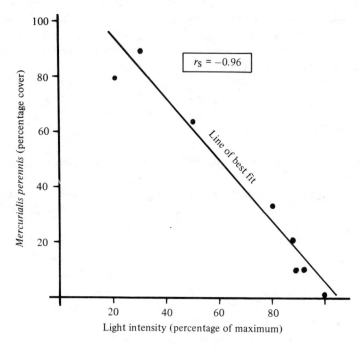

Section of transect (metres)	Light intensity (percentage maximum)	Dog's Mercury (percentage cover)
60–79	90	10
80–99	50	65
100–119	30	90
120–139	20	80
140–159	80	34
160–179	88	20
180–199	93	10
200–219	100	0

Figure 8.4 *Scatter diagram with 'best fit' line relating cover of Dog's Mercury to relative light intensity along a transect. These data refer to a part of the transect (figure 8.3) selected to investigate a gradient of light intensity. For simplicity, only section means have been plotted, but individual values (more data) would have been better (8.8)*

may be absent in the darkest places as well as from the lightest. Where there is no correlation, the points are distributed randomly (d). Most correlations are not perfect, but the points are scattered about the 'best fit' line − the greater the scatter, the weaker the correlation. A fairly good correlation is shown in (e). The scatter could be due to inaccuracies in the recording, but usually suggests that other factors are involved. Light intensity may well be an important factor with the distribution of Dog's Mercury, but, for example, local differences in soil moisture could also play a part. The line must be sloping. If it is (or nearly is) horizontal (f) or vertical, it means that all the values in one dataset are similar and not related to the variation of the other and so no correlation exists, despite the neat straight line of points. You will probably draw your lines of 'best fit' free-hand, although there is a mathematical procedure (regression) for fitting lines to points. If you are using a computer to do this, it will probably only fit them to a straight line (linear regression). Figure 8.5(c) illustrates how a computer can fit a straight line to points lying on a curve with confusing results. Figure 8.5(g) shows how a few exceptional results can distort an honest attempt to draw a 'best fit' line. It is quite permissible to ignore them, but you must point this out in your write up and justify yourself. You might find that they represented quadrats on a road crossing the transect, and you "excluded them from subsequent analysis."

Scatter diagrams enable you to

(1) display whether correlation is positive or negative
(2) identify exceptional points
(3) see whether the relationship is linear or a curve
(4) see how strong a correlation is (subjectively).

The next step is to test the significance of your suspected correlations. This is useful, but not essential, and the non-mathematical may prefer to skip to section 8.10. See flow diagram figure 8.6.

8.9 CORRELATION COEFFICIENTS

These are used to express the strength of a correlation (the degree of scatter about the 'best fit' line). They range from -1.0 (perfect negative correlation, figure 8.5a) to $+1.0$ (perfect positive correlation, figure 8.5b). A coefficient of zero indicates no correlation, and the closer the value is to $+$ or -1.0, the better the correlation. Exactly how close does it have to be before it is statistically significant (2.4(a))? This depends on how many

$r_S = +1.0$ (S**)

(a) Perfect positive
correlation

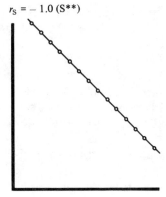

$r_S = -1.0$ (S**)

(b) Perfect negative
correlation

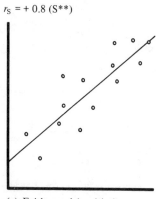

$r_S = +0.8$ (S**)

(e) Fairly good (positive)
correlation

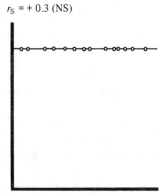

$r_S = +0.3$ (NS)

(f) No correlation (one
dataset does not vary)

$r_S = 0$

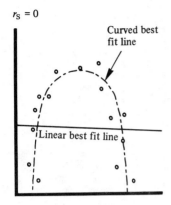

(c) Non-linear correlation
(good if regarded as curve)

$r_S = 0$

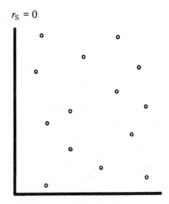

(d) No correlation
(both datasets vary
independently)

$r_S = 0.3$ (NS) with all data
$r_S = 0.9$ (S**) excluding
exceptional data

$r_S = 0.5$ (NS) with all data
$r_S = 0.9$ (S**) with irrelevant
r_S data excluded

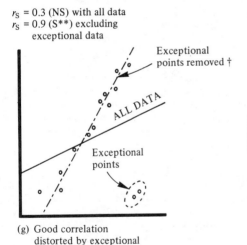

(g) Good correlation
distorted by exceptional
data

(h) Good correlation within some
of the data but distorted by
zero values in one dataset

Figure 8.5 *Interpreting scatter diagrams and correlation coefficients (8.8).* $r_S =$ *Spearman's Rank correlation coefficient. S** = Highly significant (p < 0.01), NS = Not significant. Solid line calculated by linear regression.* †*If you exclude any data the scatter diagram refers to only part of the dataset, which should be precisely identified*

147

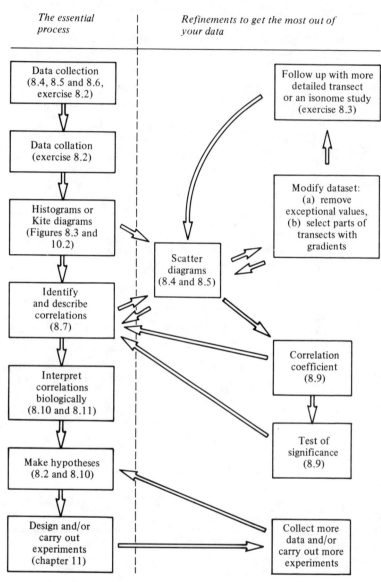

The essential process | Refinements to get the most out of your data

Data collection (8.4, 8.5 and 8.6, exercise 8.2)

Data collation (exercise 8.2)

Histograms or Kite diagrams (Figures 8.3 and 10.2)

Identify and describe correlations (8.7)

Interpret correlations biologically (8.10 and 8.11)

Make hypotheses (8.2 and 8.10)

Design and/or carry out experiments (chapter 11)

Follow up with more detailed transect or an isonome study (exercise 8.3)

Modify dataset: (a) remove exceptional values, (b) select parts of transects with gradients

Scatter diagrams (8.4 and 8.5)

Correlation coefficient (8.9)

Test of significance (8.9)

Collect more data and/or carry out more experiments

How much you do depends on the time available and how mathematical you wish your treatment to be. Do the 'essential process' at least and, if possible, some or all of the 'refinements'.

Figure 8.6 *Flow diagram showing how to deal with transect data*

data you have, and involves a table of significance levels (appendix C). If your correlation coefficient is calculated using 15 pairs of values, look up significance levels with $n - 2$ (where n = number of pairs of values) degrees of freedom $(15 - 2 = 13$ in this case). At the 5 per cent level, this is 0.514. If your calculated value (from your data) is greater than this (or less than -0.514), then it is significant. The probability of the relationship being due to chance is less than 5 per cent. Coefficients are included for each example in figure 8.5.

There are several ways of calculating correlation coefficients, but the one described here, Spearman's Rank (table 8.2) is particularly suitable for field data.

Table 8.2 Calculation of Spearman's Rank correlation coefficient

Data concern light and Dog's Mercury from figure 8.4.

Raw data		Rank order values		Difference	
Dog's Mercury (percentage cover)	Light intensity (per cent of max.)	Dog's Mercury	Light intensity	d	d^2
90	10*	6	2.5*	-3.5	12.25
50	65	3	6	3	9
30	90	2	8	6	36
20	80	1	7	6	36
80	34	4	5	1	1
88	20	5	4	-1	1
93	10*	7	2.5*	-4.5	20.25
100	0	8	1	-7	49

$\Sigma d^2 = 164.5$

Rank ordering as with Mann–Whitney test (2.6), except that the two datasets are ordered separately.

*Note how to deal with tied ranks (2.6)

$$r_S = 1 - \frac{6\Sigma d^2}{n(n^2 - 1)} = 1 - \frac{987}{8(64 - 1)} = 1 - 1.96 = -0.96$$

where r_S = Spearman's Rank correlation coefficient
n = number of pairs of values
Σ = sum of.

Warning

Correlation coefficients are very useful, but use them with care

(1) Never use one without first having done a scatter diagram and checking that the line slopes.
(2) Spearman's Rank correlation assumes that the relationship is linear (in figure 8.5c it completely failed to register a very strong correlation).
(3) It is distorted by exceptional values and by large numbers of zeros amongst the data (figure 8.5, g and h). These may mean that your coefficient comes out 'not significant' when for part of your transect it might well have been.
(4) Only use these coefficients to compare physical factors with species, not species with species (until you are experienced).
(5) Only use them to check whether a correlation that you have already detected is significant.
(6) If you find that a correlation is significant, it may well mean something biologically. If you find that it is not, this simply means that you have failed to show any significance, possibly because of deficiencies in your data (for example, there is not enough). You can still report your suspicion that correlations exist in the field, which might be shown to be significant by more rigorous sampling.

8.10 INTERPRETING CORRELATIONS

In the example provided by figure 8.5(a), it is a fact that a significant correlation exists between the data for light intensity and that for Dog's Mercury. The hypothesis mentioned above (that the species is shade tolerant) is supported by the data (assuming that the data are reliable), but not proved by it. Proof must be sought by experiment (see chapter 11), and in this case, the method described in exercise 3.2 offers one way of proceeding. If Dog's Mercury has a lower light compensation point than the species which grow in the open field but are absent from the wood, then the hypothesis would be proven, or at least strongly supported.

Hypotheses should be narrowly and precisely defined. The above example invites several hypotheses to explain the absence of Dog's Mercury from the open field. Each would require a separate experiment to test it. Examples include

150

(1) Dog's Mercury is intolerant of high light intensity.
(2) Dog's Mercury is intolerant of variations in humidity which are more extreme in the field than in the wood.
(3) Dog's Mercury is intolerant of grazing by cattle.
(4) Dog's Mercury is intolerant of trampling by cattle.
(5) Dog's Mercury is excluded from the field by competition from faster-growing species which grow in the field but are unable to flourish in the wood because there is insufficient light.

8.11 BEWARE OF 'BOGUS' CORRELATIONS

What are called 'bogus' correlations would be better described as 'potentially misleading'. Distribution of certain species of fungi, for example, may be found to be negatively correlated with light intensity on a transect crossing a woodland margin. This might invite the hypothesis that the fungi can only survive in conditions of low light intensity, but subsequent experiments would show this not to be so. Such fungi are mycorrhizal — they have a symbiotic relationship with the roots of trees. The fungus *Boletus luteus* and species of Pine (*Pinus* spp.) provide an example. In this case, the correlation between light and the fungus arises because both are correlated with a third factor, namely the distribution of trees. The trees provide the shade and also encourage the growth of the fungus. The actual correlation is not 'bogus', it is a genuine mathematical relationship.

8.12 ISONOME STUDIES

These can be used instead of transects to study environmental gradients and provide data which lend themselves to rigorous statistical analysis. In the Dog's Mercury example above, a carefully placed transect with very frequent sampling would provide a good means of investigating the gradient of light intensity on passing from the pasture, through scrub to woodland, but an isonome study would be even more thorough. An isonome study is a kind of two-dimensional transect. Instead of a single line of quadrats, a grid is used, effectively a series of parallel transects. An isonome study can also deal with a yet more complex situation in which zonation does not form neat series of belts of vegetation, but a mosaic (8.4). Examples of its use are given in chapter 9.

Exercise 8.3. Isonome study of a transition zone between two communities

What you have to decide

Read this chapter carefully, perhaps making use of the preliminary study (exercise 8.2). Your teacher may decide to omit this to save time, and make some of the decisions for you. If so, you should appreciate how much you are relying on his or her previous knowledge of the site and general ecological experience. With or without this help, you must make the following decisions, as a group

(1) The site – exactly where to lay out the grid.
(2) The species to be recorded (2.8, notes following table 8.1).
(3) The method of recording species (5.2 to 5.5). In short vegetation on fairly level ground, the method using point quadrats within a 1 × 1 or 0.5 × 0.5 m quadrat frame (5.5) could be particularly appropriate.
(4) The physical factors to measure. Use any previous knowledge that you have, such as the results of a transect in exercise 8.2. Alternatively, a few preliminary environmental measurements covering a range of factors may help. Confine yourself to getting plenty of data on a few factors that you suspect might represent gradients. A good approach is to collect the quadrat data first and return to the site when you have had a chance to discuss which physical factors may be relevant (leave the grid markers in place – inconspicuously, in case of vandalism). Collect your data objectively and accurately, and, although you are, to some extent, testing predictions, include some 'outsiders' in your range of factors. Scientific research is full of surprises!

What you need (per pair of students)

(a) A measuring tape.
(b) At least 2 marker pegs and access to a mallet.
(c) A quadrat frame (0.5 × 0.5 m or a point quadrat – see below).
(d) Recording and identification sheets.
(e) Access to (shared) equipment to measure selected physical factors (chapter 3).

Method

Each pair of students records a transect across the site as frequently as practicable according to the time available and the length of the transect. Ideally, the quadrats should be recorded contiguously – that is, without spaces between – recording every 0.5 m with a 0.5 × 0.5 m quadrat, but every 1 m will suffice. Each pair of students should also collect environmental data (and perhaps use pitfall traps (6.4(b)), or other means of sampling animals) corresponding to each quadrat. The transects of the whole class are laid parallel, and ideally at the same distance apart as the quadrat sampling interval. You must agree on a numbering system. The quadrats could be numbered from one end of the grid, and the transects given letters (for example, quadrat C5 is the fifth quadrat along the third transect).

It is quite possible that you will have decided to measure height differences as one of your physical factors (3.14). You must make your height difference measurements continuously along each transect between each sampling point, and have a value for each quadrat. Regard quadrat A1 as zero altitude, and calculate the altitude of the rest relative to it. If A2 is 15 cm lower than A1, it has an altitude of $0 - 15 = -15$ cm. If A3 is 5 cm higher than A2, its altitude is $-15 + 5 = -10$ cm. If the start of the next transect (B1) is 5 cm higher than A1, it has an altitude of $0 + 5 = +5$ cm.

Each pair of students should present the data for their transects as histograms (as in exercise 8.1 and figure 8.4). They could also analyse the data using scatter diagrams and correlation coefficients, but much more can be extracted from these data by combining the whole class's data and analysing them as a whole.

Such a 'two-dimensional transect' cannot be represented as histograms. For each species and physical factor, a map of the grid is produced. The most efficient means of doing this is to have a supply of sheets each with a copy of an empty grid, with a series of squares, one for each quadrat, with quadrats numbered along one side and letters of transects along the other. Each pair of students should collate data for one or more species and/or physical factors.

(a) Circulate your sheet(s) among the class so that everyone can transfer any relevant data that they have to them. Make sure that each sheet is clearly labelled; for example, you may have one for Dandelion (*Taraxacum officinale*) and another labelled 'soil pH'. The rest of the process is illustrated in figure 8.7.

(b) Once you have collected all the class data for Dandelion, find the highest value, and divide the range between this and zero into a number of equal-sized categories (in the example, there are six). Prepare a new sheet for Dandelion, entering the class number for each quadrat instead of the raw data. For species it is best to have the smallest category, starting at zero, but when you repeat the operation (on a separate sheet) for a physical factor (such as soil pH), range your categories from the lowest to the highest value (such as pH 4 to pH 8). You may exclude the odd quadrat with exceptionally high or low

153

values, to avoid distorting the rest, but justify yourself carefully in your write up, and indicate its position on your map (see below).

(c) Use these values on the gridded sheet to make a 'contour' map showing the density or percentage cover of Dandelion, and another to show variation in soil pH within the gridded area. This procedure is known as an 'isonome' study because points with the same ('iso-' means 'same') values are joined, or otherwise associated. You can do this by drawing lines around areas with the same value and perhaps emphasising the 'contour' map using different shading, hatching or colours (for example, for categories 1 to 6, white, yellow, blue, red, purple and black would produce a good effect). If you have altitude data, you will produce a real contour map.

(d) Trace the contours, together with the outline of the grid but no other data, on to acetate Overhead Projector (OHP) sheets (or tracing paper).

(e) You can now superimpose your contour maps on to those of other students (for other species and factors) and see which are correlated. This is particularly good using OHP sheets, as this helps class discussion.

Statistical analysis

What you have done so far is itself a very adequate means of analysing the data and looking for correlations. However, you have here a very large amount of high-quality data and this invites statistical analysis. The task may be so big that a computer will be needed to make it manageable (2.9). The approach to the analysis is similar to that for transects (summarised in the flow diagram of figure 8.7).

(a) Do scatter diagrams to investigate possible correlations between species and species, and species and physical factors.

(b) Test correlations between species and physical factors for significance, using correlation coefficients. For transects, we have suggested that you select parts of transects for calculation of the coefficients. You may similarly decide to calculate a coefficient for a defined, complete part of the grid. Perhaps one-quarter of it has a damp patch, with gradients for certain species (such as rushes and sedges) which are absent from the rest. In this case, just use the data for the damp area, but explain exactly what you did. You may also exclude exceptionally high and low values from the calculation, but justify yourself, and ask your teacher for advice if in doubt.

(c) Test the correlation between species using the chi-squared association test (5.7). Correlation coefficients would be appropriate provided that both species concerned are present in most quadrats (that is, provided

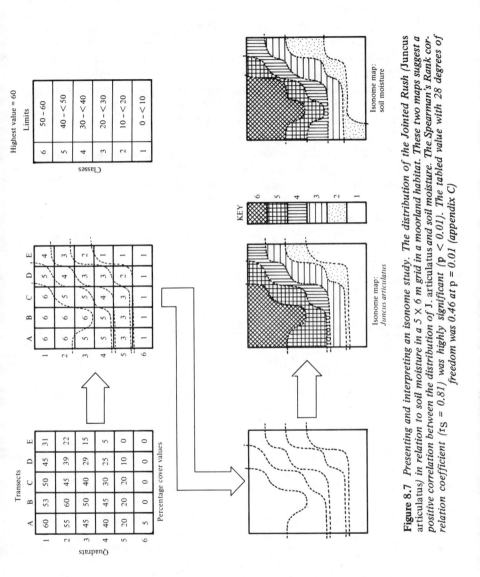

Figure 8.7 *Presenting and interpreting an isonome study. The distribution of the Jointed Rush (Juncus articulatus) in relation to soil moisture in a 5 × 6 m grid in a moorland habitat. These two maps suggest a positive correlation between the distribution of J. articulatus and soil moisture. The Spearman's Rank correlation coefficient ($r_S = 0.81$) was highly significant ($p < 0.01$). The tabled value with 28 degrees of freedom was 0.46 at p = 0.01 (appendix C)*

155

that there are only a few zeros). Again, you may need your teacher's advice.

In your write up
1. Briefly describe the site (as exercise 8.2), present your maps and describe any statistical analysis, reporting the results.
2. Discuss the extent to which the physical factors you have measured explain the distribution of species in the gridded area. Chapters 3 and 9 may help with the background information.
3. How is the distribution of one species related to that of another? For example, if certain species of rushes and sedges are concentrated in one part of the grid, it suggests that they share similar ecological niches (1.5). If they are both correlated with soil moisture, they both appear to be species of damp soil.
4. Select one or more hypotheses invited by the data. Express them precisely (8.10) and outline experiments that you would use to test them (chapter 11).
5. How would you modify your approach if you repeated this exercise?

9 Studying Particular Terrestrial Habitats

9.1 SUCCESSION, BIOTIC FACTORS AND PATTERN IN VEGETATION

In the previous chapter, it was emphasised that pattern in vegetation can often be explained in terms of physical factors, and many such factors encountered in terrestrial habitats have been dealt with in chapter 3. Two other (often related) factors, which are particularly important when considering terrestrial habitats, are biotic and successional. The biotic component of an ecosystem includes all its numerous interacting living organisms, but biotic factors which are most likely to be important in affecting the pattern of British vegetation involve direct or indirect human interference (1.7) through, for example, grazing by domestic animals, introduction of alien species such as rabbits, human recreation and forestry. Successional factors (1.6) involve time as a variable.

It is important to appreciate that zonation can involve a series of climax communities in which succession has little or no significance, but we must first consider an example in which it undoubtably has. It also illustrates the contribution of biotic factors.

9.2 SAND DUNE ZONATION – SUCCESSION AND BIOTIC FACTORS

Figure 9.1 provides a summary of the process. Sand on a beach, just above the high water-mark, is an inhospitable place for plant growth because

(1) It is often dry. Sand, especially when, as here, it lacks organic matter, has very low water retention properties, and so, as soon as it stops raining, it rapidly dries out (3.15).
(2) It is sometimes affected by seawater, at least at high tide when the sea is rough. The salts in the seawater cause a desiccating effect as a result of osmosis (3.11).
(3) It has an unstable surface which makes seedling establishment difficult.

(4) It is nutrient deficient because it lacks the organic matter which is present in most soils and provides a pool of nutrients that is continually liberated by micro-organisms. It also has low cation exchange capacity (the ability to retain cations) which, in most soils is provided by organic matter (3.16) and clay.

Although few plants can grow here, certain species such as Sea Rocket (*Cakile maritima*) and Sea sandwort (*Honkenya peploides*) can. They have xeromorphic adaptations (3.15). Such plants stabilise the sand and add organic matter (as dead leaves and roots), so improving the soil's nutrient status and moisture-holding capacity. They have brought about the first few changes which could lead to this area of foreshore becoming a forest — by succession. In many cases, Sea couch-grass (*Agropyron junceiforme*), becomes established. Small 'foredunes' of sand accumulate around the grass as stability increases, but the permanent dunes only follow once Marram Grass (*Ammophila arenaria*) and/or sometimes Sea Lyme Grass (*Elymus arenarius*) make their appearance. Marram Grass is a species which flourishes when continually buried by sand. It responds by growing vigorously, and dunes which may reach over 30 m high are the result, stabilised by the Marram grass root system, raised above the reach of waves and with an ever-increasing organic matter accumulation with ever-improving moisture retention and nutrient status. Leaching by rain starts to remove sodium chloride from developing soil. The foreshore species which helped to start the process are no longer present.

Numerous other species now become established, especially in the 'slack', the 'valleys' between the dunes where the (fresh) water-table is near or at the surface (small ponds may be present). Some of these species (such as Restharrow, *Ononis repens*) are legumes and add plant-available nitrogen to the system. The Marram Grass 'fixes' the growing mass of sand, whilst the smaller plants, including grasses like Red Fescue (*Festuca rubra*), stabilise the surface. Once the dune stops increasing in size, the Marram Grass dies out. We now have a 'fixed' dune. More species arrive. The community becomes grassland, whose species composition depends on whether the sand had a high shell content (that is, whether it is calcareous — high soil pH) or not (acid soil pH, 3.17). Such a community would be grazed by rabbits or domestic animals. Tree seedlings are very vulnerable to grazing, but the next stage, scrub, consists of 'armed shrubs', such as Bramble (*Rubus* spp.), Wild Rose (*Rosa* spp.) and Hawthorn (*Creteagus monogyna*). These are protected from grazing by their spines, and under their protection, tree seedlings become established. Often, these are Pine (*Pinus* spp.), which has xerophytic adaptations (3.15). They cast a shade which reduces the scrub species. Over a period of centuries, other tree species and a

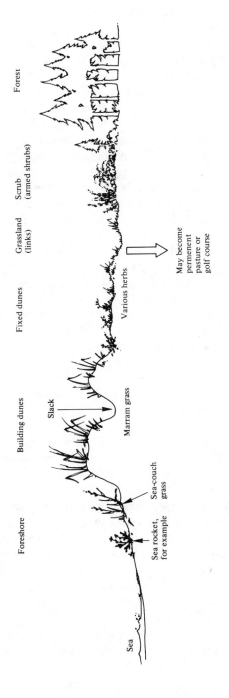

Figure 9.1 *Sand dune succession-related zonation*

variety of woodland plant and animal species will become established, and eventually, a stable climax will be reached (1.6) complete with its own self-maintaining soil system and microclimate. By this time, continued succession to seaward will have pushed back the sea several miles. The dry, sandy nature of the soil may mean Pine will remain the dominant species.

British sand dune systems rarely show the complete series of successional changes (known as a 'sere') as far as climax forest, and where they do (for example, on the Fylde coast of Lancashire), trees have been planted to speed up the stabilising process – a biotic factor (although the introduced trees may be spreading to seaward as their seedlings begin to grow in the scrub, and thus they have become a real part of the process). Some, named 'warren' on the map, may have had rabbits deliberately introduced and some are or have been used for grazing by sheep or cattle. Grazing pressure of domestic stock and introduced species like rabbits (not at equilibrium, 1.4(c)) is much greater than that of the wild herbivores of prehistoric times, and tends to suppress scrub and forest development, causing the succession to 'stick' at the grassland phase. Thus, so long as the artificial grazing pressure remains, the grassland is a permanent community, a biotic plagioclimax. Such grassland behind dunes used to be called 'links', which in many places have become golf courses. Golf links too represent a plagioclimax, maintained by mowing. Another human influence on dune systems is trampling by tourists, which may destabilise the dunes themselves.

9.3 TYPES OF SUCCESSION

Successional seres which begin in a dry habitat like sand are called 'xeroseres'. Xeroseres which start with sand are 'psamnoseres' and those starting from bare rock 'lithoseres'. Hydroseres start with open water and may result in a pond slowly becoming a marsh, and through grassland to heath, or scrub and forest (figure 9.2). You may choose to study a classic example illustrating succession, such as a sand dune system, but you may find more limited examples of it in other places. Figure 9.2 shows that succession could begin from strikingly different points, but all lead to the same climatic climax for the particular place. Most British examples include grassland or grass-heath, and below the tree-line tend towards forest via scrub. Bare soil is rare in nature, but human activity, especially in connection with arable farming, often gives rise to it. Ruderals such as Groundsel (*Senecio vulgaris*) and Shepherd's Purse (*Capsella bursa-pastoris*) are plants adapted to exploit such competition-free 'prime sites' by completing their life-cycle quickly, producing large quantities of seed and having efficient

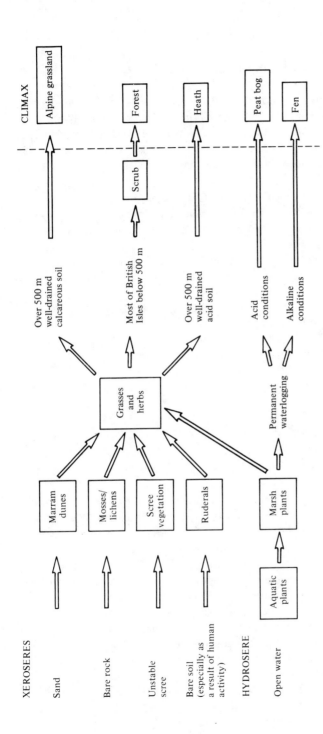

Figure 9.2 Types of succession in the British Isles. This is a very generalised picture. The 'climax' communities are the theoretical 'natural' ones. Most of the British Isles represent a biotic plagio-climax (1.6, 1.7, 9.3 and 9.4) arising from direct or indirect human interference

seed-dispersal mechanisms. If grazing pressure is reduced because a pasture ceases to be used, or because (as happened in Britain between 1952 and 1953) myxomatosis reduces rabbit populations, scrub appears, and the pasture starts to 'revert' to woodland. If some science-fiction catastrophe removed human influences from Britain, within a few hundred years, forest would probably re-establish itself in most places below 500 m.

If you have scrub in your study area, it probably implies some change in grazing regime or other (probably human) disturbance in the past 50 years, and you need to consider what it might be. Consider

(a) Ruined buildings and neglected walls, hedges or fences.
(b) Large patches of nettles – they indicate phosphate enrichment which may be associated with regular concentrations of stock, which may persist for 100 years.
(c) If there are a lot of anthills in the area, local absence of them may indicate past arable activity. Anthills often (but not always) represent over 100 years of development.
(d) Old Ordnance Survey maps. These are available for many parts of Britain from the nineteenth century.
(e) Memories of elderly local people (although anecdotical evidence should be interpreted with caution).
(f) Local place-names (for example, 'The Batts', see below).

9.4 STUDYING SUCCESSION

In your write up of a terrestrial transect study consider

(a) The only way to prove that succession is taking place is by marking out permanent quadrats and observing and recording them periodically over a number of years. To follow a sandy beach becoming forest may take hundreds of years, but invasion of pasture by scrub may be readily observed over a few years. On a Field Course, you may be able to compare your observations with records made by classes who have visited the site in previous years, provided permanent markers were left by them. It is interesting to clear a patch of vegetation near your school, college or home and observe (perhaps record quadrats) over several years, to see succession happening.
(b) Look for evidence of former vegetation under the existing community. You may be able to find remains of Marram Grass by digging under 'links' grassland (not, of course, possible on a golf course!).

(c) If you already have good reason to believe succession is taking place (as you would on a sand dune system), a study of zonation may enable you to describe it. Succession involves changes in time at a particular point, but as these are progressive, the zonation will afford a spatial representation of its course. As you pass from the shore across dunes to the distant woodland, the terrestrial vegetation that you walk over gets older with every step. In fact, the age of the community is positively correlated with distance from the sea. If your transect crosses a woodland margin through scrub into pasture, possibly succession is taking place, and the woodland, led by a belt of scrub, is advancing into the pasture. If this is true, you can suggest which species of herbs and grasses accompany the early stages of scrub formation, the sequence in which they are lost as shade increases, and whether some woodland species invade the scrub in advance of the trees.

(d) If you cannot find physical factors to explain why one part of the transect is scrub and another is woodland, consider possible succession in relation to historical events (human or otherwise).

Example

One of the authors studied a small area of woodland and scrub, with open grassy clearings (rich in interesting plants and butterflies), between marshy ground and a river. The name of the site (The Batts) was found to mean 'an island in a river' in Middle English. This, and a seventeenth century map, confirmed that the marshy area was an old river bed, and so succession was (or had been) taking place in the shape of a hydrosere and perhaps was likely to continue until it became woodland, indistinguishable from the rest. Patches of scrub differed markedly in the sizes (and presumably the ages) of the Hawthorns of which they consisted. It was suspected that this was related to changes in stocking density and to myxomatosis outbreaks in the past. The clearings were a conservation priority, and it was suggested that they were being encroached by the scrub and that steps to cut it back were necessary. A permanent transect produced no evidence of successional change over several years, and it was found that scrub density was negatively correlated with soil depth. In fact, the clearings were stable climax communities maintained by shallow soil, apparently the summits of old river shingle banks. The zonation in scrub density did not imply succession. In a site nearby, with similar clearings amongst Hawthorn, the situation looked the same but was not. The clearings had ariseen because, in the past, the scrub had been cut as fuel for a now disused lime-kiln. Successional changes threatened the rare plants, as scrub invaded the clearings. Scrub clearance was needed to conserve one site, but not the other.

9.5 SELECTING SITES AND DECIDING HOW TO STUDY THEM

In chapter 8, three approaches to relating vegetation and animal populations to environmental factors were described, which are summarised below.

(1) Comparing two (or more) homogeneous sites (exercise 8.1).
(2) Transects to study zonation in relation to environmental gradients in a general sense, or to examine in detail a gradual transition from one zone to another (exercise 8.2).
(3) Isonome studies to consider mosaics or to examine transition zones in detail (exercise 8.3).

The rest of this chapter is designed to help you select sites, and decide which of the three methods to use, which environmental factors are likely to be worth measuring (chapter 3) and which animals to sample (chapter 6). It is not intended to be anything like a complete survey of suitable habitats, but provides examples and advice to illustrate principles which can be applied much more widely. Specific background information is included but you will need additional sources. *Britain's Green Mantle* by Sir Arthur Tansley (revised by Proctor, 1968) published by George Allen and Unwin, will be particularly helpful, along with many of the 'New Naturalist' series published by Collins.

9.6 SITES TO COMPARE (SEE EXERCISE 8.1)

(a) Two different types of woodland
Particularly recommended: Deciduous *versus* Evergreen, Beech (dense shade) *versus* Ash (less shade), or managed *versus* unmanaged woodland (but with the same dominant species). These studies can be readily combined with exercises 6.2, 6.3, 6.4, 6.5, 6.6 and 6.7.

Important additional background is an account of the woodland management.

(b) North-facing and south-facing slopes
Ideally, a small steep-sided valley with opposite slopes of similar gradient running east to west and of uniform underlying geology is required. Chalk and limestone valleys are very suitable. This study is readily combined with exercise 6.1.

Measure especially the soil temperature (3.10) over as much of the day as possible, but you need to consider other environmental factors (for

164

example, wind, slope, soil pH, soil moisture and soil organic matter), to check for other possible differences. A difference in soil temperature should be expected, but you need to consider how other factors may be related to it.

Important background information includes any differences in grazing regime between the opposite slopes.

(c) Two areas of grassland

The areas of grassland should be fairly close (similar climatic conditions), perhaps parts of the same field. Particularly recommended are sheep *versus* cattle grazing, heavily *versus* lightly grazed, well-drained *versus* marshy, well-cultivated *versus* rough pasture, golf-course green *versus* 'rough' (check that the course is not in use!), 'cricket square' *versus* the margins of the pitch (get the groundsman's permission!), and newly established *versus* old lawn. In acid uplands of Britain, the dominant grass is often the Mat Grass (*Nardus stricta*) where it is thought to have replaced heath following overgrazing in the past. It is unpalatable to stock. Another type of upland grass which sheep prefer is *Agrostis-Festuca*, dominated by Fescues and Bent grasses. These two communities are worth comparing. On calcareous soils, under-grazing leads to *Agrostis-Festuca* being replaced by 'coarse' grasses like the Upright Brome (*Bromus erecta*).

If, for example, you consider that you are studying well-drained *versus* marshy parts of a field, any attempts to explain a vegetation difference in terms of soil moisture assumes that other factors are constant. You need to check this by considering a variety of possibilities, as well as measuring soil moisture.

Important background information is present and past land use (for example, information on grazing, mowing, fertiliser treatment, drainage, and use of pesticides). If you suspect that a difference in grazing is important, try to get precise information from the farmer, such as number of sheep per acre/hectare (1 hectare = 10 000 m^2 = 2.4 acres), and rates of fertiliser and lime application per acre (many British farmers think in terms of hundredweights per acre − 1 hundredweight = 50.8 kg).

(d) Calcareous/non-calcareous geological boundary

Calcareous rocks include chalk and limestone, and you may be able to find places where they have a boundary with non-calcareous rocks like sandstone and granite by inspecting a geological map. This boundary may be covered by glacial drift or river deposits (alluvium), and so do a few preliminary soil pH determinations on either side of where you think the boundary is. If there is not a clear difference, then this is not a suitable site. You would also expect to see an obvious difference in vegetation

165

(calcicoles and calcifuges, 3.17). Establish both your grids on either side of any transition zone, where the vegetation is homogeneous. If your objective is to consider the effects of the geology on the vegetation, your two sites need to be as similar as possible in other respects, so avoid if possible obvious differences in slope, drainage, land use etc. Measure soil moisture, soil organic matter and soil temperature, as well as the obvious soil pH values.

If the area is being grazed during your visit, estimate the density (subjectively) of animals on both sides of the boundary (not, in this case confining yourself to the sampling grid). Numbers of rabbit droppings in your quadrats will indicate rabbit preferences. In some cases, the distribution of nests of certain bird species may be related to the boundary too. Perhaps combine your vegetation sampling with an Earthworm study (exercise 6.1).

Important background information is the present and past land use. Calcareous soil tends to provide better grazing than 'acid' soil, and may have been more cultivated and grazed in the past. Look for traces of an old wall or fence along the boundary.

9.7 USING TRANSECTS TO STUDY ZONATION (SEE EXERCISE 8.2)

As mentioned in chapter 8, the seashore (chapter 10) is particularly recommended for this type of study, but some suitable terrestrial situations are given below. In selecting environmental factors to measure, the object is to select those existing as gradients which may help to explain zonation, not those which are constant throughout the transect.

(a) Sand dunes (9.3)
Important possible environmental gradients: soil moisture, slope/altitude, soil temperature, wind speed, and, if woodland is included, humidity, air temperature and light intensity. Some soil pH measurements are suggested.

Note that section 9.3 describes a very generalised picture. Your site may differ in many matters of detail. Such a transect will be very long (perhaps over a km). Use 1 x 1 m quadrats every 10 m (or more), with the whole class working as a team.

Important background information is land use of 'links' area, tourist pressure on dunes, any management of wood, or history of tree planting.

(b) Pond margin (open water-rushes/reeds-marsh-pasture) – and perhaps contine to scrub and woodland (figure 9.2)
Important possible environmental gradients: soil moisture, humidity,

altitude, and soil temperature. Include other measurements in (a) if continued to woodland.

Important background information is whether pond is dredged out periodically (if so, how frequently), any fertiliser or effluent which may drain into pond (chapter 7), use of adjacent land, and age of pond.

Animal data would be a complex study in itself, whose form would depend on the site (chapters 6 and 7).

(c) Pasture-scrub-woodland (example in chapter 8)
Important possible environmental gradients: light intensity, humidity, wind speed, air and soil temperature, and soil moisture.

Consider whether succession may be important with respect to the scrub area (see especially section 9.4). If the transition zone is compact, an isonome study could be better than a transect. Most of the animal studies in chapter 6 could be combined with collecting the vegetation data.

Important background information is present and past use of the pasture, including, if possible, stocking density, fertiliser and herbicide treatment, and whether field is mown for hay or silage (if so, when?). Also management of woodland for timber and/or shooting and any use for recreation by the public.

(d) Effect of altitude
For this you need a mountainside which offers you a fairly uninterrupted slope rising from near sea-level to a summit over 500 m. Your safety must be an essential consideration. Measure slope and altitude (if possible), and most, if not all the other factors considered in chapter 3 could be worth measuring (8.4).

(e) Effect of slope
This does not require a high mountain, but you need a hillside, or part of a mountainside with different gradients, including some very steep parts, preferably with scree (fairly bare ground with loose stones). Again safety aspects must be taken seriously. Start or finish on level ground.

Important possible environmental gradients: slope (much more important than altitude), soil depth, soil moisture content, soil pH (look for evidence of leaching and redeposition of bases like calcium ions).

Important background information includes the geology, present and past land use, and tourist pressure (slopes are vulnerable).

Consider the possibility of successional colonisation and stabilisation of scree.

(f) Transitional zones

If in any of the situations mentioned in (9.6) a preliminary investigation shows a considerable transition zone rather than an abrupt change, you may prefer to concentrate on this instead of on homogeneous vegetation on either side, and do a transect or isonome study. For details consult (9.6).

9.8 ISONOME STUDIES (SEE EXERCISE 8.3)

As mentioned above, isonome studies can often be used to study transition zones and their associated environmental gradients. Not only are they a sounder basis for statistical analysis, but individual students can do one of a series of closely parallel transects and combine them to form an isonome study. They can also be used to study mosaics (8.4) and other situations where zonation is small-scale and not in regular belts. A few suggestions are given below.

(a) A damp patch in a field or lawn, including some of the adjacent drier ground.

(b) A field with mediaeval ridge and furrow pattern. Try soil moisture and levelling. Sometimes different species of buttercups tell a story. Although these ridges and furrows are thought to be connected with peasants ploughing the same strip every year with paths in between, nobody knows why they persist.

(c) Effect of trampling on pasture: usually people do not keep strictly to a path, and so there is a gradient of trampling as you approach it. Have the path across the middle of your grid. Perhaps you could use distance from the path as a factor for correlation. Measure soil depth and moisture content. Can you somehow measure the degree of soil compaction?

(d) Light and shade in a wood: select an area of woodland floor under a broken canopy with an obvious pattern of light and shade. Light intensity is best measured over a whole day (3.6). If you cannot do this, you could use the bicarbonate indicator method (3.7) with a shade plant like Dog's Mercury, examining the tubes at hourly intervals throughout the day. You might find that bare patches of soil on the woodland floor are explained when you discover they have so little light that even Dog's Mercury is rarely above its compensation point. Relating the density of tree seedlings to light intensity may also be interesting. In a natural Ash wood, for example, the seeds lie

dormant until a tree falls down, breaking the canopy and letting light reach the ground.
(e) Hummocks and hollows in a peat bog: the hollows are wet, and vegetation grows more rapidly there than on the drier hummocks, whose vegetation is very different. The hollows (involving succession) become hummocks and *vice versa*, and thus a 'raised bog' grows. You need a small quadrat (say 0.25 x 0.25 m) and you may have difficulty in making it lie flat (improvise!). Measure soil moisture and pH, and level it (exercise 3.8).
(f) A small stream with a marshy margin: lay out the grid so that the stream passes through the middle, and try to include drier ground on one or both sides. Measure especially soil moisture and temperature, and level the grid (exercise 3.8). Consider whether land use differs on opposite sides of the stream – for example, perhaps one side is a meadow with cows and the other is ungrazed (and untrampled) near a hedge (in which case consider light and humidity) or a road.

Example
A group of students used a 5 x 17 m grid on ground which sloped through open woodland to a stream with marshy margins in a Gloucestershire nature reserve. They recorded only three species, two species of Golden Saxifrage, *Chrysosplenium oppositifolium* and *C. alternifolium*, and the Lesser Celandine (*Ranunculus ficaria*), using point quadrats per unit area (5.5). They also measured soil moisture (3.15) for each 1 x 1 m grid square. Both species of *Chrysosplenium* were positively correlated with soil moisture, but *C. oppositifolium* had a more significant correlation coefficient, indicating that it was restricted to the wettest places. *C. alternifolium* had the highest cover values in the wettest places, but seemed to have a greater tolerance of moisture stress, and also grew in drier places. *R. ficaria* was negatively correlated with soil moisture (and with both the other species). It grew in the drier parts of the grid. The slope was levelled. The relationship between soil moisture and slope was not unexpected. The flat area near the stream was the wettest. The above correlations were visually evident from the isonome diagrams (like figure 8.7), which were drawn by computer (authors' 'ecostat' program). Scatter diagrams (8.8) and correlation coefficients (8.9) provided confirmation.

Warning
In all scientific work, you must make hypotheses yet beware of preconceptions. In a hydrosere, for example, you expect there to be a gradient in soil moisture. This will account for much of the zonation, but there

could be other gradients; for example, if there is a tree overshadowing part of a transect there will be a light intensity gradient whose effects on zonation will be unconnected with the hydrosere.

10 Seashore Habitats

This chapter calls on a great deal of information that is presented in chapters 4, 5, 6, 8 and 9. It deals with seashore habitats as a separate topic because this type of environment has certain distinctive features that are not encountered elsewhere.

10.1 THE SHORELINE

The key gradient is the degree of exposure to air which is conveniently correlated with distance from the low tide level, and so environmental gradients are more predictable than on land.

The shoreline is divided into five different zones as follows

(1) Splash zone
(2) Littoral fringe – upper shore
(3) Eulittoral zone – middle shore
(4) Upper sublittoral zone – lower shore } Littoral zone
(5) Sublittoral fringe

These divisions are defined according to tide levels during the year, but they should not be regarded as rigid and the size of the zones can vary considerably owing to a variety of factors such as degree of exposure of the shore; there is also a direct relationship between the slope of the shore and the widths of the zones. Figure 10.1 shows how the zones may be associated with tide levels.

10.2 PROBLEMS FACING SHORE ANIMALS AND THEIR EFFECTS ON ZONATION

The factors that affect the animal and plant life on the seashore inevitably have the effect of restricting certain species to different regions of the shoreline.

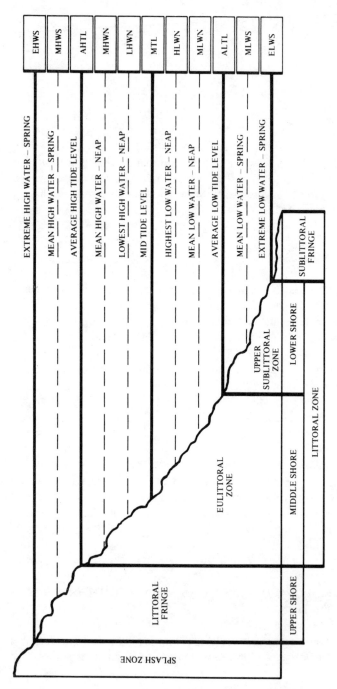

Figure 10.1 *Zonation on a rocky shore in relation to standard tidal levels*

(a) Water loss

One of the main factors influencing animal and plant distribution is how long they have to remain exposed to the air while the tide is out. What makes this a particularly important factor is that seashore animals are essentially aquatic and so are liable to water loss by evaporation and eventually death from dehydration. So to be able to survive in periods of water shortage they must be well adapted to restricting water loss.

Water loss can be fatal either because of a change in the internal environment as a result of an increase in the body fluids (this affects the osmotic potential of the body, which may assist desiccation as the cells lose water by osmosis to their surrounding fluid), or because of a lack of oxygen to the cells because of a lack of water flowing over the surface of gills (so reducing gas exchange). Aquatic animals may suffocate even if they are surrounded by oxygen-rich air because their gills will stick together. Water is also lost by excretion; many marine animals excrete ammonia which is highly toxic and must be diluted with large quantities of water. This is a serious problem on the seashore where water loss is to be avoided at all costs.

Certain species that are found up the shoreline have developed ways of surviving these conditions by restricting water loss. Algae, for example, are covered in a layer of mucilage, while most surface animals have a shell that can be closed in various ways if they are uncovered. Some species of gastropod excrete uric acid which is less toxic and so requires little water for removal.

Two species are very well adapted to life at high levels up the seashore. The first is a seaweed, *Pelvetia canaliculata* (channel wrack) which is olive-brown in colour; it is tolerant of desiccation and in those months when tides are low it will become dried up, brittle and almost black. The second is a species of barnacle, *Chthalamus montagui*, which may extend right up into terrestrial regions, it is particularly well adapted to withstanding desiccation.

(b) Wave action and degree of exposure

Certain species are better adapted to withstanding strong wave action, and it is these that dominate exposed shorelines. The problem facing animals on exposed shores is the destructive impact imposed by stones and other materials that are thrown against them. Animals that live on the surface are in danger of being crushed, while burrowing individuals are likely to be churned up by the waves. Continuous wave action makes it difficult for individuals to become attached, and those that do are often torn off. It is for these reasons that very exposed shores may be completely barren of plant and animal life, while those that are slightly less exposed are populated by barnacles and mussels. Barnacle shells are firmly cemented to the rocks, while the common mussel, *Mytilus edulis*, has byssus threads that penetrate deeply into the rocks. As the degree of exposure decreases other species appear in the environment, such as the common limpet, *Patella vulgata*, which has a powerful foot for attachment, together with large algae that also have strong holdfasts.

(c) Changes in temperature

When the tide is out there are sometimes wide fluctuations in temperature on the shore. Increases in temperature are due to strong sunshine heating up the water left behind on the beach, causing it to evaporate; this is accompanied by an increase in the salinity of the water remaining. Intertidal organisms may also be subjected to low temperatures caused by snow, ice or frost, and as the ice moves it crushes the organisms.

The main problems associated with wide temperature fluctuations include the effect on the enzymes in the body, freezing of tissues and, if an increase in salinity is also present, the organisms may encounter osmotic problems. Those organisms that do not have adaptations enabling them to survive such fluctuations are only found in those regions of the shore which are uncovered briefly or not at all (that is, the sublittoral fringe).

(d) Seasonal changes

Depending on the time of year, various animals may move to a different zone because of changes in temperature, so zonation can vary during the year.

(e) Changes in light intensity

At low tide the shore is exposed to direct sunlight and the accompanying problems of dehydration, increased temperatures etc. However, it is possibly the decrease in light intensity when the tide is in that limits the distribution of certain algae. In bright light the rate of photosynthesis increases, particularly in algae and rock pools, and this increases the oxygen level of the water and there is a corresponding increase in pH.

(f) Predation

Shore animals are subjected to marine predators when they are submerged, and at low tide they are preyed upon by terrestrial and aerial animals. Those organisms that are exposed at low tide are therefore preyed on by two different sets of predators, while those that remain submerged are at slightly less risk. However, there is, of course, a lot of predation of fish by seabirds.

For example, the distribution of algae may be limited because of grazing by herbivores, while the distribution of encrusted animals such as mussels may be limited by carnivores.

(g) Immersion

The majority of shore animals use the oxygen that is dissolved in the water for respiratory purposes, but a few are air breathing. Those individuals that are air breathing are killed by prolonged immersion and so the distribution of these species is restricted to the higher levels. In addition, water pressure varies with submergence and so sensitivity to changes in water pressure is also important.

The small periwinkle, and the rough periwinkle, have a reduced ctenidium and the mantle cavity is modified to form a lung, so a higher percentage of these is found in the littoral fringe and splash zone.

(h) Behaviour

Many animals can detect a change in their surroundings and react accordingly, because of their kinetic and tactic responses. There are many behavioural modifications displayed by animals to overcome the changes in their environment, such as an increase in temperature or salinity at low tide. For example, if *Patella* is splashed with freshwater it pulls its shell down hard and remains still, but if it is repeatedly splashed with seawater it begins to wander about, so it is able to protect itself from changes in salinity in its surrounding water.

10.3 EXAMINING ZONATION ON ROCKY SHORES

A quick glance at a rocky shore will be sufficient to notice that there are distinct bands running across it. On closer examination it can be seen that these bands are composed of strips of different species of seaweeds. Closer investigation reveals that there are also different species of animals present in these different zones.

Rocky shores are the most permanent of all shores. Because of the conditions that they are exposed to, only the most secure and hardest rocks will be able to stay on the shore, and the seaweed which usually covers them, except in the most exposed places, will protect them from sea and air.

The surface of the rocks is a very important factor in determining the type and quantity of animals that are present. If the rock is smooth, then it is more difficult for organisms to attach themselves, but if the rocks are rough and uneven, then attachment will be easier. Similarly, the gradient of the shore is important. If the shore is flat then there will probably be a lower density of animals, while if it is sloping there may be many crevices and overhangs that will contain life.

Before carrying out any investigation on the shore, it is advisable to obtain information about the tides at the particular place you intend to study. The easiest way to do this is to obtain a copy of tide tables for the area. Using this, it is possible to find out exactly when the tide is due to go out and come in, and so you can avoid being cut off by the tide. Once this information has been obtained, an investigation can be carried out in safety.

It is also useful to obtain as much background information about the area as possible, such as the presence of nearby factories or power stations, sewage discharge, oil spillage from accidents at sea, and, if there are any freshwater streams running on to the beach, find out what the land they pass through is used for. The factors are important as they may have dramatic effects on the flora and fauna.

Exercise 10.1. Examining zonation using a belt transect

The methods used here are described in chapters 3, 5 and 8 in more detail. Read exercise 8.2 before proceeding.

What you need
(a) Several measuring tapes.
(b) White chalk sticks.
(c) A wooden peg (optional).
(d) A 0.5 × 0.5 m quadrat.
(e) Record sheets.
(f) A suitable key (one purpose-made by the teacher for the shore being studied is easier for first-timers to use).
(g) Equipment for measuring physical factors (chapters 3 and 4).

Method
The tape should be placed as near to the last terrestrial plant as possible, or as high up the shoreline as you intend to sample. This point is marked with a wooden peg, if the substrate is soft, or if not a chalk mark can be made on the rock and labelled 0 metres. This should be done when the tide is on the way out. The tape should then be laid out down the beach to as near the low tide level as possible, and marked at intervals, of say 5 metres, or however frequently you intend to sample.

Physical measurements can be made from the point of origin of the tape down towards the sea by two groups of students, one group measuring slope (3.14) and the other temperature (3.8). Other factors that might be worth measuring include light (exercise 3.1) and humidity/evaporation rates (exercise 3.7).

Working in groups of two or three, the remainder of the party can concentrate on recording the flora and fauna of the shore. It is best to lay the first quadrat at the point nearest to the sea because as the quadrats are studied and recorded, the tide will be chasing you instead of coming towards you rapidly. It is important to work quickly and accurately as time is limited.

The types of shore life vary greatly, and range from lichens, sponges and seaweeds to molluscs, crustacea and even small fish. It is important that you look in all possible places in your quadrat and do not restrict yourself to what you can see easily. So remember to look under stones, no matter how small, in cracks and crevices, under and on seaweeds, as well as on the rock surfaces.

It is possible to record all the species you find, but this is time-consuming and inaccurate identification is likely, so it is better to restrict identification to particular species that illustrate zonation well. Probably the best thing to do for a first study is to restrict recordings to a few indicator species of seaweed and animals, or simply to record only the seaweeds (table 10.1); the latter technique is elementary and restricts any further studies, and should not be used if follow up work is intended. Recordings of percentage cover or the actual numbers of individuals can be made. In many instances you may have a total percentage cover of more than 100 per cent (why?).

Handling the data
When all the data have been collected, they must be collated back at the laboratory. With these data, histograms (figure 8.3) or kite diagrams (figure 10.2) of the distribution of each species can be drawn. As measurements of light intensity and slope have been recorded too, you may wish to see if there is a correlation (8.7 to 8.11) between altitude (directly related to length of exposure by the tide) or light and a particular species. Some species of the *Rhodophyceae* (red algae) are absent from areas of high illumination, so it is valuable to make light recordings in the same places as the quadrats.

In your write up
In addition to the points mentioned in exercise 8.2, you should also explain the following.

177

1. Discuss which organisms are dominant in the different zones along the transect. How are they adapted to living in their particular zones?
2. Does there seem to be certain species that occur together? How could you account for this?

Table 10.1 An example of zonation of some seaweeds on a semi-exposed rocky shore

Seaweed zone	Species present	Tide
Pelvetia	Enteromorpha compressa Pelvetia canaliculata	
Fucus spiralis	P. canaliculata Cladophora rupestris Ulva lactuca Catenella repens Ceramium rubrum Chondrus crispus Corallina officinalis	– – EHWS – – –
Ascophyllum nodosm	Cladophora rupestris C. flexuosa Fucus vesiculosus F. spiralis A. nodosm Pilayella littoralis Polysiphonia lanasa Cladostephus verticillatus Laurencia pinnatifida Chondrus crispus	– – LHWN – – –
	Corallina officinalis Furcellaria fastigiata Fucus serratus F. vesiculosus	——— MTL ———
Fucus vesiculosus	Lamentaria articularia Ascophyllum nodosm Ulva lactuca	– – HLWN – – –
Laminaria	Laurencia pinnatifida Lomentaria articulata Rhodymenia palmata F. serratus Himanthalia elongata Laminaria spp.	– – MLWN – – –
		ELWS

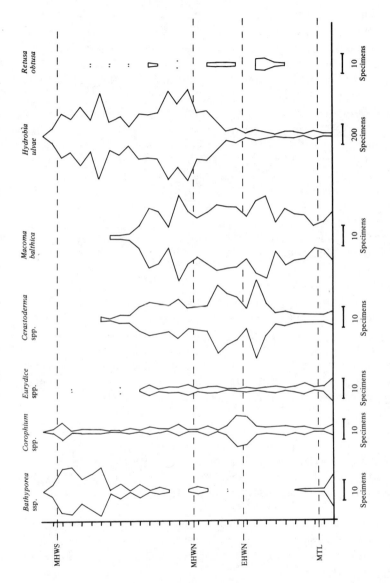

Figure 10.2 *Kite diagram to show the intertidal distribution of animals in an estuary*

179

10.4 ALTERNATIVE WAYS OF RECORDING OR FOLLOWING UP INVESTIGATIONS

(1) You may simply wish to find out where particular species of barnacles are found along a shoreline, in which case you need only to record 'present' or 'absent' on your record sheet. In addition you may wish to estimate the percentage cover of the barnacles to see how their populations increase and decrease along the shore. This method can be used for other species too, such as the periwinkles.

(2) Alternatively, after doing a complete transect up a beach, the results you obtain may lead you to think that two species always occur together — that is, there is a positive correlation between them. To check this you could carry out another transect to see if this is indeed the case, and then go on to test your hypothesis using the chi-squared test for association between organisms (5.7).

(3) The effects of predation can also be studied using a transect by recording the presence and absence of two species such as barnacles and dogwhelks. In each quadrat where both are present you must record three things

(a) the number of living barnacles
(b) the number of dead barnacles (that is, those with no opercular plates)
(c) the number of dogwhelks.

You may also record, if you wish, quadrats that have no dogwhelks and very few, if any, living barnacles. If there are a large number of quadrats that have lots of dead barnacles but there are relatively few dogwhelks around, what would you suggest might be the cause of this?

(4) You may wish to relate your study to the degree of exposure of the shore. To do this you need to have a copy of a biologically defined exposure scale, such as Ballantine. When carrying out an exposure study it is important to record the slope of the beach and to relate this to the abundance of indicator species. There may also be differences in the sizes of the different zones.

(5) As an alternative to a full study on the degree of exposure, the effect of wave action on just one species, such as *Fucus vesiculosus* (bladder wrack) can be studied. There are several features that can be studied on this species, such as

(a) the diameter of the holdfast
(b) the circumference of the stalk (stipe)

(c) the number of bladders
(d) the amount of branching
(e) the length of the thalli.

There is no need to study every quadrat to do this study. It is sufficient to take measurements from one sample in each zone. Is there a significant change in any of the measurements?

(6) An isonome study could be carried out (8.12, exercise 8.3) on a rocky shore, and a particularly good place to do one is on a shore which has a freshwater stream running on to it. This is because, where fresh and salt water mix, there is an area of brackish water which may support slightly different life.

10.5 ROCK POOLS

These are well worth studying in detail if time is available, as they contain a wealth of plant and animal life. By their very nature they are not permanent structures, and often they are completely covered by the incoming tide which washes away the animals they contain, leaving behind different individuals as it recedes. Others, higher up the shore, may be left uncovered by the tide for days or even weeks, and these are subjected to great extremes of conditions; for example, rainfall will decrease the salinity and temperature of the water, while hot sunshine will cause evaporation, which in turn will increase salinity and temperature. Therefore, life found in rock pools is subjected to wide fluctuations in environmental conditions. Obviously, therefore, the position of the pools on the beach is an important factor to consider when carrying out an investigation.

Exercise 10.2. Measuring physical and chemical factors in a rock pool

Refer to chapter 4 for details of the methods.

What you need
(a) A measuring tape.
(b) A metre rule.
(c) A comparator.
(d) A temperature probe.
(e) A conductivity meter and probe.
(f) Quantab 1177 chloride titration papers.
(g) An oxygen meter and electrode.

(h) A pH meter and an electrode.

(i) Record sheets.

Method

Carry out the investigation when the tide has gone out, you will then be able to select a rock pool close to the sea, and make recordings from it before the tide turns.

Select a suitable rock pool; a small one is a good choice, as close to the sea as possible. Measure the dimensions of the pool and draw a scale map of it; do not forget to measure the depth in different places too. Indicate all the dimensions on the map, and sketch on any over-hanging rocks and areas of seaweed. Following the instructions given in chapter 4, take measurements of oxygen (4.2), pH (exercises 3.12 and 3.13, section 4.3), conductivity (4.6), salinity (4.8) and temperature (4.4, exercise 3.4).

To make the exercise comparative, select two more pools, one in the middle of the beach and another as high up the beach as you can, and take the same measurements from these. You will then be able to compare pools from different parts of the beach and see if there is in fact any difference between them.

It is also a good idea to study one pool over a 24 hour period. It is not necessary to have an expensive data recorder, although it is useful to be able to make recordings automatically over the required period. Besides, unless you have access to a private beach, it is not wise to leave expensive equipment lying about. It is quite satisfactory to simply take measurements at equal intervals over the 24 hour period — say at 7 a.m., 10 a.m., 1 p.m., 4 p.m., 7 p.m., 10 p.m., midnight (and 3 a.m. if anyone is enthusiastic enough to get up!). It is important to select a pool that is high enough up the beach so that it will not be covered at high tide. Also, make sure that there is easy access to it, especially if it is to be visited after dark.

In your write up

1. Compare the readings you get for each rock pool. Can you account for any differences?

2. If a 24 hour study is undertaken, plot the results as a series of graphs. Study the fluctuations of the parameters over time. How do you account for these changes?

Exercise 10.3. Making a quantitative study of the flora and fauna of a rock pool

Refer to chapter 6 for further details.

What you need
(a) A hand net.
(b) A bucket.
(c) A white tray.
(d) A handlens.
(e) Record sheets.
(f) A suitable key.

Method

This should be carried out in the same pool as the physical and chemical measurements are taken from. If for some reason or another this is not possible, then make a map of the pool as described in exercise 10.2. First of all, study the rocks at the sides of the pool, and the water/rock margin. Count and record the animals and plants you identify. Then, using the hand net, collect samples of water and swimming animals. Place each netful in the white tray, examine your catch, and count and record what you find. Then place the contents of the tray in the bucket so that you do not catch and count the same animals twice. Repeat the process until you fail to catch anything. Before you return your catch to the pool, look closely at the bottom and sides of the pool and record any sedentary animals you see there, such as *Actinia equina* (beadlet anemone). Do not forget to look under all the seaweed and in the cracks and crevices too.

In your write up
1. Why do you think there is a sharp boundary between the sides of the pool and below the waterline?
2. Which animals are found above and below the waterline?
3. Are there any organisms which are found only in the water? If so, can you account for this?
4. What gives many rock pools (although perhaps not the one you have studied) a pink colouration?
5. If you examined the same pool in a week's time you would not expect to find the same numbers of mobile animals, while the quantity of seaweeds and sedentary animals would remain more or less the same. Why?

10.6 EXAMINING ZONATION ON SANDY OR MUDDY SHORES

At first sight a stretch of sandy shore looks as if it does not have any life in it, let alone show any zonation. However, it is a minefield of burrowing animals and possesses a wealth of planktonic animals and plants. Sandy shores are made up of very fine particles of hard rock, usually quartz.

Because of the small size of the particles (they range between 2.0 mm and 0.2 mm) water is held by capillary action in the small spaces between the sand grains. This prevents them from rubbing against each other and allows animals to live in the water between them when the tide has gone out.

10.7 FACTORS AFFECTING ANIMAL LIFE

(a) Desiccation is rarely a problem on a sandy shore, except at the upper-most zone, the littoral fringe (or strandline). The strandline is the region occupied by masses of debris which is thrown up by the sea. Exactly where it is varies according to the position of the high tide level over the year. At this level the sand is left uncovered for some time and the sand dries out, and air penetrates between the particles. The animals found at this level are so well adapted that they are practically terrestrial; for example, *Talitrus saltator* (the sand hopper) is an air breather. There may be as many as 25 000 of these per square metre of sand. Therefore, unlike a rocky shore, where degree of exposure to desiccation is a major factor in zonation, a sandy shore is not so dependent on it.

(b) Wave action is a constant problem, and as the sand is continually churned up it is not a suitable substrate for attachment. For this reason, there are very few attached organisms visible, except on the more sheltered stretches of sand. Here, wave action is reduced and organisms can attach themselves to embedded stones etc.

Wave action is the most important factor in controlling the distribution of the sand-dwelling organisms, because it influences, either directly or indirectly, several important factors, namely stability, particle size, slope, oxygen content and organic content.

In very exposed regions there may be no life in the sand, because none can withstand the powerful crushing forces of sea against sand. As the degree of exposure decreases, then certain individuals that do not need a permanent burrow can live between the sand grains, but are strong enough to withstand the pressures of sand movement. One such is *Tellina fabula* (*Tellinid*).

On the more sheltered beaches where the sand is more stable, popu-lations of animals that live in permanent burrows can survive – for example, *Sabellaria* spp. More sheltered sand tends to have a larger amount of organic matter which provides food for a large number of small inverte-brates that live between the grains. There may also be a dark region a little way below the surface that indicates the presence of sulphide bacteria that are anaerobic. The burrowing animals then have to rely on obtaining their oxygen from the water by pushing siphons or tentacles up into the water, and they must have some way of storing oxygen when the tide is out.

The most sheltered stretches tend to have a lot of small particles of sand mixed with a lot of clay particles, giving the sand a darker, muddy appearance and texture. This is the most stable shore with very little wave action, making it ideal for animals such as *Cerastoderma edule* (cockle) or *Arenicola* spp. (lugworm) and *Mya arenaria* (sand gaper).

(c) Stability, slope and water content of sandy shores are all dependent on particle size. Sand falls into two main categories, dilatent and thixotropic.

Dilatency is the term given to sand which, when you walk on it, forms a white 'halo' around the foot. This is caused by the foot applying pressure to the sediment which alters the packing of the particles and so the interstitial spaces become distorted, and the water there is insufficient to fill the new spaces, so the sand looks white. If greater pressure were applied, it would be met by increased resistance to penetration. So a burrowing worm or bivalve would find it difficult to move through. Therefore, dilatency means that burrowing would require the expenditure of a lot of energy.

Thixotropy is the term given to sand where the application of pressure does not meet with an increase in resistance, and there is no whitening at all. If pressure is applied there is distortion, but there is sufficient water present (20-25 per cent) to fill the spaces. So, pressure application is met by a decreased resistance to penetration (this is what happens in quicksand).

Areas of sand tend to be some way between these extremes. It is likely that burrowing animals would be found in dilatent rather than thixotropic sand, which goes very soft and wet under pressure.

(d) Temperature and salinity are two other factors which are important to animal life. Both these factors are hardly affected by changes on the surface sand which may be caused by hot sunshine or rainfall.

Exercise 10.4. Examining zonation along a belt transect

Refer to exercise 8.2 before proceeding.

What you need
(a) Measuring tapes.
(b) Wooden pegs.
(c) A mallet.
(d) A 0.25 x 0.25 m quadrat.

(e) A large spade.

(f) A bucket (for carrying water).

(g) A bowl (for holding dug-out sand).

(h) A sieve.

(i) A suitable key.

(j) Record sheets.

(k) Plastic bags (for sand samples).

(l) Waterproof pens.

Method

Place the first peg at the last terrestrial plant or at the start of the strand-line. Lay the tape out as far down the beach as you wish to go, and mark out with wooden pegs every 5 m or how frequently you intend to sample.

Working in groups of two or three and starting at the end of the tape nearer to the sea, record what you find in each quadrat. This is not so straightforward as recording species on a rocky shore, because this time you must dig for them. Lay the quadrat flat on the sand and count and record; that is, record the density (as with daisies in exercise 2.1) of any living animals that may be on the surface. Then dig to a depth of 0.5 m over the area of the quadrat (that is, 0.5 m^2). Place the sand in the bowl, and then collect all the animals, by washing a little of the sand through the sieve at a time. Identify, count and record all the buried animals that you have decided to deal with along the transect. Take a sample of the sand and label it with the distance along the transect from which it was taken (for example, 65–65.25 m) for analysis back in the laboratory (exercise 10.3).

Handling the data

As described in exercise 8.2, the class results must be collated before any conclusions can be made. The density values for each species present in each quadrat must be written on a master record sheet, and this then made available for the whole group.

The results can be plotted as histograms, but it is useful to plot all the information on a single sheet of paper as a series of kite diagrams. Kite diagrams are drawn with a single pair of axes: distance along transect/number of specimens. A standard length to represent say 10 specimens is decided upon before any data are plotted. Dealing with one species at a time, find the largest number of individuals that you will have to plot and mark this on the diagram in its correct position. Draw a line in the middle of these two points, parallel to the y-axis, very lightly with a pencil. This line will help to guide your other points. Then plot all the other values you have, ensuring that your faint pencil line lies in the middle of the two points plotted for each quadrat. When you have plotted all the points for each species, using a ruler, join up all the points on each side of the faint pencil line, and rub out the central line. You may if you wish shade in the strip of graph paper you have marked out, figure 10.2.

In your write up
1. Discuss the distribution of different species along your transect. Does there seem to be any zonation? If so, can you account for the fact that certain species are found in some places and not others.
2. As you were working along the transect, did you notice any marks on the surface of the sand which after a while you realised indicated the presence of particular animals in the sand? What were the marks like and which animals did they indicate the presence of?

Exercise 10.5. Examining the moisture content of sand

Refer to exercise 3.10 for moisture content determination.

In your write up
1. How does amount of sand moisture relate to the distribution of animals along your transect?
2. Can you explain why some animals appear to be more numerous when the sand is wetter, while others decrease in numbers?

What you need
(a) Sand samples collected from each quadrat along the transect correctly labelled.
(b) Apparatus as in exercise 3.16.

Method
Carry out the procedure described for soil samples (exercise 3.16).

In your write up
1. Why is organic matter important in the sand?
2. What is it used for?
3. Does there seem to be a decrease or increase in animals where there is a lot of organic carbon? Why?

Exercise 10.6. Examining the organic matter content of sand

Lugworms (*Arenicolidae*) are filter feeders and live in U-shaped tubes in sand, one end of which draws water in (inhalant) while the other passes water and sand particles out (exhalant). Their presence in the sand can be detected by large casts of sand at the exhalent end of the tube at low tide, while at the inhalent end there is a small depression visible.

What you need
(a) Measuring tapes.
(b) A bucket.
(c) A spade.

Method
The first thing to do is to mark out an area on the sand, in the area covered by casts, with some measuring tapes. The size of the area studied may depend on a number of things, namely, time available, number of visible casts, and the distribution of the casts. Once a suitable area has been marked out, count the number of casts visible, and record the number. Then, find the inhalant end of the tube and place the spade parallel to the cast and depression and dig until you find a worm. Place the worms in a bowl and, when you have dug up all the worms you can find, count all the ones that are alive. Return all the worms and sand to the area.

In your write up
1. How did your initial count of the number of casts compare to the actual number of worms you found?
2. Why might an investigation restricted to counting the number of casts lead to an under-estimate of the population of lugworms in the area studied?

10.8 ESTUARIES

An estuary is basically a stretch of sea water that has extended into a low-lying valley that contains a freshwater river. Where the fresh and salt water mix there is a decrease in salinity and the water is called 'brackish'. Animals that live in estuaries must be able to tolerate variations in salinity over 24 hours, as the tide goes in and out and fresh and salt water mix. Clearly, the animals that are found in the river itself are freshwater organisms, while those found in salt water are marine organisms, but there is a wide band between fresh and salt water where the animals are neither one thing nor another, and are especially able to tolerate extremes. At high tide, sea water dominates the estuary, while at low tide, especially if it is raining as well, freshwater dominates. See table 10.2.

Most marine invertebrates have body fluids that are at the same osmotic concentration as the surrounding sea water; these animals do not have to cope with the excessive absorption of water by osmosis, as do those of

188

Table 10.2 An example of the distribution of some animals in an estuary

SEA

Mouth of estuary	*Semibalanus balanoides*
	Chthalamus montagui
	Patella vulgata
	Littorind spp.
	Mytilus edulis
	Cerastoderma edule
	Nucella lapillus
	Crangon crangon
	Arenicola marina
	Carcinus maenas

(the distances these extend up the estuary depend on their powers of osmoregulation and any protective coverings or burrowing power that they have)

Mid estuary	*Enteromorpha intestinalis*
	Fucus ceranoides
	Corophium volutator
	Hydrobia ulvae
	Nereis diversicolor
	Scrobularia plana
	Macoma balthica
	Carcinus maenas
	Sphaeroma rugicauda
	Eurydice pulchra
	Gammarus zaddachi
	Whitebait (herrings and sprats)
Upper estuary	*Palaemonetes varians*
	Chironomus
	Tubifex
	Notonecta

Figure 10.2 shows the distribution of certain species of animals along an estuary.

freshwater (for example, starfish) and these are called 'osmoconformers'. Animals that regulate their body fluids to an osmotic concentration above the surrounding medium (for example, *Carcinus* (shore crab)) are called 'hyperosmotic'. Most osmoregulators are not capable of moving into and inhabiting brackish water, although some can tolerate it for short periods. Those animals that are able to osmoregulate only over a narrow range of concentrations (for example, *Maia* (spider crab)), are called 'stenohaline', while those that have become adapted to life in estuaries and have the ability to osmoregulate over wide ranges (for example *Carcinus* (shore crab)) are called 'euryhaline'.

189

Some species can avoid the problem by burrowing down into the sand or mud at low tide. For example, *Arenicola* has very limited osmoregulatory powers, but by burrowing down as far as 20 cm or more it can avoid changes in salinity.

The mud of estuaries has been brought down from the land in the river, and contains a large amount of organic debris, and it may also contain pollutants from factories and farms. The build-up of silt and debris at the mouth of the river together with debris brought in at high tide may lead to the formation of mud flats beneath calm shallow water. These waters may experience wider fluctuations in temperature. The lowest temperatures usually coincide with the lowest salinities as increased rainfall in the winter months will add to the freshwater outfall, and *vice versa*.

Populations of estuaries face strange situations, and in the middle of the estuary only the most euryhaline forms can survive. Sample an estuary at different points between its mouth and where it becomes a non-tidal river, using similar methods to those used in the exercises in this chapter. All the sampling points may be over 100 m apart; this would be a transect study in which salinity would be the principal gradient. What other environmental factors might be worth measuring?

11 Experimental Ecology

11.1 WHAT IS THE PURPOSE OF EXPERIMENTS?

Much of what you do in field work concerns making observations. In the course of this you look for relationships between the distribution of species (both plants and animals) and environmental factors. In chapter 8 (especially 8.2, 8.10 and 8.11), it was emphasised that correlations based on observations suggest hypotheses but do not prove anything more than that a relationship exists. An experiment is the next stage in the scientific process, and this is the way that hypotheses are tested.

Consider this example. The density of Duckweed (*Lemna minor*) was found to be statistically greater in one of two streams. One hypothesis considered was that the two streams differed in flow rate. Duckweed only grows in stagnant water, or slow-flowing water-courses. When this was investigated (4.11), both streams were found to have a similar (slow) flow rate, and so that hypothesis was rejected. Another hypothesis about the 'low Duckweed' stream being shadier was rejected once light intensity measurements (exercise 3.1) were made over one day. A reasonable hypothesis could be that something to do with the water might explain the difference, but you can never be sure that there were not other differences in factors that you did not measure because you lacked the necessary equipment, or that you simply never thought of. Perhaps some weedkiller entered the low-Duckweed stream several weeks ago which has long since been washed away or, unknown to you, another party of students had been using nets in the stream a few days before your visit.

Any field situation is so complicated that you need to simplify it and bring it under control. The first step is to find out about the species you intend to use. Section 11.8 discusses Duckweed and its use in experiments.

11.2 CONTROLS

Experiments are more correctly called 'controlled experiments'. Conditions are kept as constant as possible. In a University laboratory, plants could be grown in a growth chamber in which light intensity and constant or varied in a controlled way to simulate night and day. You may not have one of these, but you could use a greenhouse, perhaps with electric lights, to provide as controlled an environment as you can. Even with the most sophisticated equipment, you can never be sure that you have total control, nor that you have thought of every possible variable. This problem is overcome using a control experiment. This is exactly the same as the experiment itself, except for one variable, the one under consideration. Other conditions outside your control will still vary (in a greenhouse, light and temperature), but the 'control' will experience them too. By comparing the result of the 'experiment' with that in the 'control', we can identify the effect of the factor that has been manipulated by the experimenter.

In the example, the variable is the source of the water, and the hypothesis is that the water in the high-Duckweed stream is more conducive to the growth of Duckweed than water from the other stream. You must use Duckweed collected from one of the two streams because Duckweed from somewhere else may be different genetically. The Duckweed in the two study streams may also differ genetically, but this is a separate hypothesis to be set aside for the present (to be reconsidered if the present hypothesis has to be rejected). Use low-Duckweed stream water in the 'experiment' and high-Duckweed stream water in the 'control'. See (11.8) for further practical details.

11.3 COLLECTING EXPERIMENTAL DATA OBJECTIVELY

Growth of Duckweed can be easily made numerical by putting three fronds in each flask initially and then counting the number of fronds (11.9), after they have had chance to multiply, but before they have filled the flask (usually 1-2 weeks). Growth of other plants and animals is usually measured by weight increase (11.9), and most other forms of data can take the form of numbers too.

11.4 THE NEED FOR REPLICATION AND A STATISTICAL APPROACH

In an experiment you aim to have one controlled variable and you expect to be able to attribute any variation in the results to this. Variation from other sources affecting the 'control' and the 'experiment' differently would complicate the experiment. Two such sources which are to some extent unavoidable are

192

(1) Experimental error. Even if you are very careful, you cannot avoid these entirely, and, in any case, equipment you are using may not be completely accurate.

(2) Biological variation. In chemistry, one gram of copper(II) sulphate is the same as another, but this is not the case where biological material is concerned. No two Duckweed fronds, woodlice or even two peas in a pod are identical. Using a clone of asexually produced material (which is genetically uniform), like most Duckweed populations, does minimise one source of this variation, but that originating from the environment remains.

As with comparing randomly collected quadrats (2.4), the issue is not simply "is there a difference between the 'experiment' and the 'control'?," but "is the difference statistically significant?." For this reason, biological experiments usually involve replicates. In the Duckweed example, instead of two flasks, one with low-Duckweed stream water and the other the 'control', several (say 10) flasks are used for each. At the end of the experiment, the mean number of fronds for each group are compared. The result may be so obvious that no further statistics are necessary, but two datasets of 10 numbers could be tested for a significant difference using the Mann–Whitney method (2.4) although if you have met the Student's t-test in a statistics course, it would be appropriate with this kind of data, even though it is parametric (2.5).

A point to consider with replicates is that replication itself causes variation. Perhaps the flasks with Duckweed furthest away from the glass of the greenhouse received less light, or those near the door experienced more temperature fluctuation. If all your 'controls' were near the windows or door, they may differ from the 'experiment' for the wrong reason. The solution is to randomize them. Making sure you have labelled them first, mix them up, and the significance test will effectively take such variation into account (although it may make significance more difficult to establish). Randomization is best done by numbering the flasks and then arranging them in order with random number tables (2.2).

11.5 EXPERIMENTS MUST BE DESIGNED PREDICTIVELY

No scientist carries out an experiment without having an idea of the result. A properly designed experiment can have only one of two possible overall results. Either one thing happens, and the hypothesis is proved, or something else happens and it is not. The null hypothesis principle applies (2.4). Design your experiment such that you eliminate the possibility of difference being due to chance variation, before declaring your own hypothesis proved.

11.6 ONE EXPERIMENT USUALLY LEADS TO ANOTHER

You make a lot of assumptions when you design an experiment and if you get a significant result, you should add "within the conditions of the experiment", which should be carefully described. The same is true if it turns out not to be significant. If the low-Duckweed stream water in the experiment did not affect growth, perhaps it would have if only you had kept the greenhouse at a lower temperature. The only way to check this is by further experiments.

Suppose the low-Duckweed stream water did have less growth of Duckweed than that of the 'control' when the experiment was carried out. All you have achieved is to narrow down the problem. The next question is "what is it about the low-Duckweed water which is the main cause of the effect?." You return to the streams and carry out water analysis (4.7). The low-Duckweed water is found to have significantly more nitrate, nitrite and phosphate, but less dissolved oxygen (4.2).

These chemical data provide several hypotheses, each of which needs a separate experiment. You may need to try them all, but you use previous knowledge to select the most likely. Surely nitrate and phosphate are plant nutrients, and these should encourage growth, not reduce it? Green plants produce oxygen (in the light) so why should reduced oxygen affect them? When you read in a book that nitrite can be toxic in quite low concentration it begins to look worth an experiment. Use mineral medium instead of stream water (11.8). This contains no nitrite unless you add it yourself. Add nitrite to it, at various concentrations, with a 'control' without added nitrite. Since you analysed the water, include the actual concentration at the time of your visit as one of the treatments. As it may vary from day to day, try some treatments above and below it. If it was measured as 25 ppm (parts per million) try 0 (control), 10, 20, 30, 40 and 50. Note that this experiment is more complicated than the last which had only two treatments (low-Duckweed water and a 'control'). You should still use replicates for each treatment, but as 60 flasks might be too many to handle, you have (say) 5 instead of 10. If you had no idea which concentration to use, do a pilot experiment with levels of 0, 1, 10, 100 and 1000 ppm to find the range to use. Here, the prediction if your hypothesis is true is that there will be a negative correlation between growth and nitrite concentration (8.4 and 8.5) and a correlation coefficient could be tested for significance. The results of this experiment and your water analysis help to

194

answer the second question — "is there enough nitrite present to significantly reduce growth?".

11.7 SUMMARY OF THE DUCKWEED EXPERIMENT

Hypothesis: the chemical nature of the low-Duckweed stream water has an adverse effect on the growth of Duckweed (*Lemna minor*).

There were two treatments, one using low-Duckweed stream water, and a 'control' using high-Duckweed water, with ten replicates of each treatment. Each replicate consisted of 50 cm^3 of water in a 250 cm^3 conical flask fitted with a cotton-wool plug (figure 11.1). Three fronds of Duckweed collected from above the outfall were placed in each flask. They were placed under a bank of fluorescent lights in a greenhouse (in which the temperature varied during the experiment between 10 and 30°C), arranged in a randomized fashion. After 17 days the numbers of fronds in each flask were determined, and the mean calculated for each treatment. The two sets of data were also tested for significance using the Mann–Whitney test.

The results confirmed the hypothesis, and led to further hypotheses about what aspect of the low-Duckweed water's chemical composition produced the effect. A further experiment confirmed that nitrite did have a toxic effect on Duckweed at the levels actually detected when the water was analysed. These findings may not, of course, apply in similar situations.

The next question is "what was the source of the nitrite?". As nitrite is a nitrogen cycle intermediate, decaying organic matter is a possible origin. Another visit to the site pointed the finger of suspicion at a pig unit beside the low-Duckweed stream, but this invited a study of the water upstream and downstream of it, involving yet more experiments. A scientist's work is never done!

Exercise 11.1. Carrying out an experimental study

The rest of this chapter gives further information which might help in designing your project, and a few more general suggestions.

(a) Make hypotheses precisely and specifically (8.10). It helps in formulating a hypothesis to write it down. Beware of vague hypotheses which are really a confused complex of ideas. Well-defined hypotheses seem simple, brief and rarely contain words like 'and' and 'or'.
(b) Is the hypothesis testable? You must be realistic about this. Are suitable methods, materials and equipment available to you? Is the species

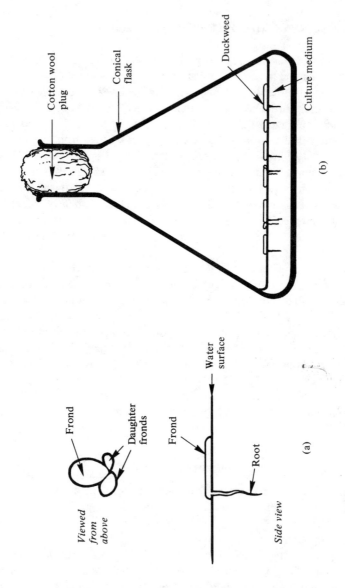

Viewed from above

Frond

Daughter fronds

Frond

Water surface

Root

Side view

(a)

Cotton wool plug

Conical flask

Duckweed

Culture medium

(b)

Figure 11.1 *Duckweed as an experimental organism: (a) external features, (b) culturing Duckweed*

you plan to use easily maintained in a laboratory or greenhouse? Hypotheses involving historical events are not testable by experiment, but the historian's data may help to confirm them (8.2 and 9.4).

(c) Which statistical methods do you intend to use (thinking of this as you plan the investigation often helps experimental design).

(d) What do you predict will be the result if your hypothesis is true, and what would happen if it is not? If you realise you are not sure, then something is wrong. You are now beyond the 'let's just see what happens' stage.

(e) Is a pilot experiment needed to decide the necessary conditions, such as temperature or nutrient concentration?

(f) Are conditions as constant as possible? A common problem, for example, occurs when using different amounts of light because electric lights and the sun also produce heat, and by varying light you may unwittingly also vary the temperature. Immersing all your treatments in the same large volume of water offers one possible solution.

When you have presented and analysed your results, consider them in the light of your field data and those of others (that is, consult books) and let them lead you to further experiments.

In your write up

1. Describe all methods so accurately that others could repeat what you did after reading your account. State very clearly the hypothesis that each experiment was designed to test.

2. Present your results in such a way that they are easily interpreted. Use tables, graphs, scatter diagrams, statistical tests, diagrams and photographs where necessary.

3. Write a discussion of your methods and results. Mention any weaknesses and improvements in techniques and experimental design you would use if you could start again. Be specific. Discuss how you would extend the project if you had time.

4. Briefly summarise your conclusions.

11.8 DUCKWEED AS AN EXPERIMENTAL ORGANISM

This species (figure 11.1), especially the commonest British species, *Lemna minor*, is a useful subject for growth experiments. Each individual consists of a frond which floats on the surface of water and a root which extends vertically downwards (these may fall off when they are exposed to toxic substances). They grow rapidly in suitable conditions and, as they

197

do so, produce new fronds, which eventually separate from the parent, providing a form of asexual or vegetative reproduction. Growth can be readily recorded by knowing how many you had to start with, and counting the number after a known length of time.

The fronds can be grown in conical flasks containing a mineral medium (figure 11.1). In the example, stream water was used but, in many experiments commercially available water culture media (such as Hoagland's or Long Ashton) can be used at one-tenth the recommended concentration. To find the effects of, for example, different levels of phosphate, use the 'without phosphate' version, and add known amounts of phosphate as Na_2HPO_4 or NaH_2PO_4. Alternatively, you could use 'complete culture' and add known amounts of substances to test their toxicity as in the example (11.6).

When collected from the wild, Duckweed is usually contaminated with algae, which also flourish in the medium, providing an unwelcome complication. This can be overcome by shaking fronds for different lengths of time (from 0.5 to 10 minutes) in 10 per cent sodium hypochlorite solution (use freshly made solution – it deteriorates on storage), rinsing in distilled water and putting them into separate, freshly prepared flasks. The idea is to kill the algae but not the Duckweed. Use the flasks in which the plants recover, but which are algal-free, genetically uniform cultures (clones) as material for experiments. As the stock culture flasks get full, sub-culture by transferring a few fronds to fresh flasks.

11.9 MEASURING GROWTH

This is an important basis for many experiments related to ecology, in which hypotheses concern certain conditions being favourable or unfavourable to the growth of a particular species. Usually you will be studying plants. The same principles apply to animals (for example, Hopkins' famous experiments on vitamins using rats), but often you will not be able to use enough for adequate replication, or will feel it is unethical to experiment with them.

Duckweed (11.8) provides a useful simple system but usually growth of plants is measured by dry weight increment. You could, for example, grow the common Rye Grass (*Lolium perenne*) in pots containing soil from a common source, or in water culture, and measure dry weight increment under different conditions. After allowing them to grow for, say, three weeks, harvest the leaves and shoots, place them in labelled paper bags and dry overnight in an oven at $100°C$, and weigh (giving 'dry weight'). If you weigh the replicates separately, you can test the results for significance by the Mann-Whitney test (2.6) or the *t*-test. Most accurate results are obtained by germinating seeds in a seed tray, and planting the same number of seedlings in each pot. The problem is that you need to know the initial dry weight, but finding this involves killing the plants. The solution is to estimate the initial dry weight of the seedlings by taking a sample of them and finding the average dry weight.

11.10 FIELD EXPERIMENTS

Laboratory or greenhouse experiments are important in elucidating specific problems, but in ecology, the best place to test hypotheses is often in the field. A complete project may involve an interplay of field observations, laboratory experiments and field experiments.

You may want to test the hypothesis that a patch of scrub represents a stage in succession. In the example discussed in (9.4), such a hypothesis was disproved by marking out a transect permanently with pegs, recording it and waiting several years before recording it again. The first set of data represented a 'control', time is the controlled variable and subsequently collected data acts as the 'experiment'. Another experiment carried out in the same habitat tested the hypothesis that the vegetation was affected by large grazing animals (probably rabbits). Quadrats were permanently marked out, and carefully recorded. Rabbits and other large animals (for example, deer) were excluded by erecting a wire mesh fence around some of them (that is, they were made into exclosures). Other quadrats were left without wire mesh, open to grazing, and acted as 'controls'. One year later, the quadrats were re-recorded, and several significant differences were found between the exclosures and the 'controls'. Thus the hypothesis was proved. The results provided details. Many of the small herbs had disappeared, perhaps because the much taller grass (mean height 62cm in the exclosures with 4 cm in the 'controls') shaded them out. This new hypothesis could be tested by growing mixtures of the grasses and herb species of the site in greenhouse experiments, and studying competition between them in which you simulate various grazing regimes using scissors. Alternatively, you could use plots on the site in which you release known numbers of rabbits.

No seedlings of scrub plants were found amongst the tall grass, perhaps for the same reason. The experiment had also shown that it was not grazing which kept down scrub in this habitat. The negative correlation between soil depth and scrub density suggested a new hypothesis, but this was more difficult to test by experiment. How would you proceed next?

A common type of field experiment is to mark out sample plots, record them by quadrat methods and add different types of nutrients such as phosphate at various levels, preferably in the growing season (usually late spring). There should be enough plots to allow replication, perhaps in a randomised fashion. Several months (or years) later, the results are collected in the form of more quadrat data.

11.11 INVESTIGATING TOLERANCE OF MOISTURE STRESS

Moisture stress is a common problem in terrestrial habitats and on the seashore, and adaptation to it can be a feature controlling the distribution of plant species (3.15, chapter 10).

Suppose, for example, you want to test the hypothesis that a variety of seaweed species occupies particular places on a seashore because it differs in its ability to tolerate water loss. Collect samples of, for example, *Fucus serratus, Pelvetia canaliculatus, Fucus spiralis, Ascophyllum nodosm, Fucus vesiculosus* and *Laminaria* spp. Weigh several samples of each species. Lay each one out under a lamp or electric fan for one hour. Next weigh all the samples again, and record the result. Repeat this process every hour for a minimum of 5 hours. Calculate the average percentage water loss for each species. Plot your results for each species on the same graph. You could test some of the differences for statistical significance (2.6). This will show you how the species differ in their ability to retain water, and see if this helps to explain their position on the seashore.

You could similarly compare terrestrial plant species suspected of differing in their moisture stress tolerance (3.15), modifying familiar 'O' level methods for studying transpiration. These involve potometers (either the simple type, or the Philip Harris electronic one which can be interfaced or used with a datamemory, 2.10), or weighing leaves, hanging them in front of a fan and reweighing them. Percentage weight loss indicates water loss, and you would expect this to be less the more xeromorphic the species. Again, if you have adequate replication, tests of significance could be applied.

11.12 BEHAVIOUR EXPERIMENTS

These provide an ethically satisfactory opportunity to experiment with animals, especially if you use invertebrates. The techniques using woodlice in choice chambers in most 'O' and 'A' level textbooks can be modified for many small invertebrates met with in your pitfall trapping (6.2(b)). You may already have met habitat preference experiments for woodlice involving humidity and light. You could extend this to other species (such as centipedes) but consider hypotheses suggested by your own observations of their distribution in the field. You may have found that their distribution is correlated with certain plant species, and hypotheses about food preferences could be tested.

Suppose that you have noticed that the Freshwater Shrimp (*Gammarus* spp.) is most abundant in streams with a stony bed, and you hypothesise that they have a preference for a stream bed consisting of particles of a certain size. You need a set of choice chambers containing stream or pond water, in which each chamber has a different layer on the bottom (sand, small pebbles < 0.5 cm, pebbles 0.5-1.0 cm and pebbles > 1.0 cm). Place about 20 *Gammarus* in the central chamber, leave for 30 minutes, and record the numbers in each chamber. Repeat at least five times, calculate means and test for significance.

Finally, an example to illustrate some of these points.

Exercise 11.2 To examine the hypothesis that when *Hydrobia ulvae* (mud snail) emerges from the mud surface at high tide, it does so to feed

What you need
(a) An aquarium.
(b) 200 *Hydrobia ulvae* (at least).
(c) 5 'dirty' sticks, preferably ones that were picked up on the beach when *Hydrobia* was collected or left stuck in the sand for several days.
(d) 5 'clean' sticks.
(e) A supply of sea water.
(f) Enough mud to cover the bottom of the aquarium.

Method
Scrape the 'dirt' off the clean stick and examine under a microscope to find which organisms it contains. Cover the bottom of the aquarium with the mud you have collected; there should be enough for a depth of 2.0 cm. Place the wooden sticks arranged in a randomized fashion, with one end in the mud and the other against the side of the aquarium. Place the *Hydrobia* on the mud and allow sufficient time for the snails to bury themselves. Next gently pour in the sea water (to simulate high tide), and leave for at least 2 hours. Record the number of snails on the sticks and calculate the average for clean and dirty ones. Carry out a Mann-Whitney test for significance (2.6).

201

In your write up
1. Is the hypothesis proved?
2. What further hypotheses do your data invite?
3. Comment on the design of the experiment. How would you improve it if you did it again?

Appendix A: 200 Pairs of Random Coordinates between 1 and 30

21 12	1 15	1 18	16 24	11 23	16 25	28 27	10 18
27 14	3 2	30 14	9 16	29 6	30 27	22 17	9 12
17 9	11 10	24 4	1 11	13 21	14 13	13 10	23 7
5 12	18 30	3 4	6 28	3 7	9 7	24 8	9 15
3 15	13 23	6 14	27 5	5 10	30 18	22 19	26 24
25 3	12 28	18 19	3 19	19 20	13 17	19 7	1 14
4 25	3 13	6 3	26 11	2 10	30 10	16 28	10 22
20 9	5 22	18 14	17 21	17 3	9 19	6 30	30 8
1 25	3 7	2 27	13 14	22 26	20 13	6 23	11 21
6 26	19 15	3 16	20 8	13 23	9 26	15 7	30 26
26 9	2 29	5 22	17 7	2 14	9 20	17 9	9 30
21 4	11 19	27 26	13 23	4 30	28 21	19 14	18 28
28 19	9 16	14 21	1 7	11 10	16 21	30 9	11 19
13 15	5 11	8 6	23 26	2 14	7 15	8 7	11 7
8 23	6 29	8 12	17 6	24 22	17 5	27 5	8 23
24 26	28 25	1 15	27 23	10 16	5 12	19 8	18 11
14 10	24 17	7 7	3 12	30 21	8 26	2 28	14 28
21 10	6 8	19 13	6 20	24 3	30 27	14 8	28 18
30 20	28 7	17 4	7 18	23 12	28 18	15 9	20 26
21 11	2 23	6 12	17 5	26 27	23 16	16 13	7 15
3 15	18 19	20 10	30 2	30 3	9 27	19 5	23 18
18 11	19 10	2 11	3 2	30 19	5 30	24 9	27 21
14 26	18 9	3 5	5 3	2 11	13 25	5 21	28 24
11 2	6 11	7 7	25 11	5 19	2 25	5 16	1 7
25 21	24 3	26 3	15 14	29 15	16 24	28 18	21 5

Notes

1. Use these numbers in sequence, either downwards or across, backwards or forwards.
2. Avoid always starting in the same place. When you get to the end (or beginning), start at the beginning (or end) again. If a group of people use this table independently, working on the same grid, they should use different parts of the table.
3. You can use any column(s) or line(s) as an individual series of random numbers (rather than coordinates) for other purposes (for example, randomization of replicates, 11.4).
4. You will find longer tables like this, with a bigger range of numbers, in many statistical textbooks.
5. These random numbers were generated by a Sinclair ZX Spectrum microcomputer with the command 'RANDOMIZE O' (see manual) preceding the 'RND' statement in the program.

Appendix B: Significance Levels for Mann−Whitney Test

n	5 per cent	1 per cent	n	5 per cent	1 per cent
4	11	9	33	952	904
5	18	15	34	1013	962
6	26	22	35	1075	1022
7	37	32	36	1139	1084
8	49	43	37	1206	1148
9	63	56	38	1274	1214
10	79	70	39	1344	1282
11	96	87	40	1416	1351
12	116	105	41	1490	1423
13	137	125	42	1565	1496
14	160	146	43	1643	1571
15	185	170	44	1723	1648
16	211	195	45	1804	1727
17	240	222	46	1888	1808
18	271	251	47	1973	1891
19	303	282	48	2060	1975
20	337	314	49	2149	2062
21	373	348	50	2240	2150
22	411	385	55	2724	2620
23	451	423	60	3256	3138
24	492	462	65	3836	3703
25	536	504	70	4464	4315
26	581	548	75	5141	4976
27	629	593	80	5865	5683
28	678	640	85	6638	6439
29	729	689	90	7459	7243
30	782	740	95	8329	8094
31	837	793	100	9247	8994
32	894	847			

n = the number of pieces of data in *EACH* dataset (assuming that both datasets contain the same number of values).

Notes
1. For there to be a significant difference between datasets, the calculated value ($T1$) must be *LESS* than the tabled value (note that this is the opposite way round to most tests, such as correlation coefficients).
2. This table assumes $n_1 = n_2$; that is, both datasets contain the same number of values. It is suggested that you always plan your investigations so that this is so. Some statistics textbooks have tables for when n_1 and n_2 differ, but note that they rarely deal with values of $n_1 > 15$ nor $n_2 > 28$.

Appendix C: Spearman's Rank Correlation Coefficients (r_s) at the 5 per cent and 1 per cent Levels of Significance

Degrees of freedom (n − 2)	5 per cent	1 per cent	Degrees of freedom (n − 2)	5 per cent	1 per cent
2*	none	none	24	0.388	0.496
3*	1.000	none	25	0.381	0.487
4*	0.886	1.00	26	0.374	0.478
5*	0.750	0.893	27	0.367	0.470
6*	0.714	0.857	28	0.361	0.463
7*	0.683	0.833	29	0.355	0.456
8*	0.648	0.794	30	0.349	0.449
9	0.602	0.735	35	0.325	0.418
10	0.576	0.708	40	0.304	0.393
11	0.553	0.684	45	0.288	0.372
12	0.532	0.661	50	0.273	0.354
13	0.514	0.641	60	0.250	0.325
14	0.497	0.623	70	0.232	0.302
15	0.482	0.606	80	0.217	0.283
16	0.468	0.590	90	0.205	0.267
17	0.456	0.575	100	0.195	0.254
18	0.444	0.561	125	0.174	0.228
19	0.433	0.549	150	0.159	0.208
20	0.423	0.537	200	0.138	0.181
21	0.413	0.526	300	0.113	0.148
22	0.404	0.515	400	0.098	0.128
23	0.396	0.505	500	0.088	0.155

n = the number of pieces of data in *EACH* dataset.

To be significant, the calculated value for r_S must be greater than the tabled value.

*Values for degrees of freedom 2 to 8 (inclusive) are specific to the Spearman's Rank procedure (calculated by Kendall). The rest are as used for the ordinary parametric (2.5) correlation coefficient (r).

Values for 2 to 8 (inclusive) degrees of freedom are reproduced by permission of the publishers, Charles Griffin & Company Ltd, of High Wycombe, from M. G. Kendall, *Rank Correlation Methods*, 4th Edn, 1970. The rest of the table is reprinted by permission from *Statistical Methods*, Seventh Edition, by George W. Snedecor and William G. Cochran (c) 1980 by the Iowa State University Press, Ames, Iowa, USA.

Index